Mass Media in Agricultural Extension

Mass Media in Agricultural Extension

B. Lahiri
S.D. Mukhopadhyay
M.M. Adhikary

2012
DAYA PUBLISHING HOUSE®
New Delhi - 110 002

ISBN 9789351241027

Published by : **Daya Publishing House®**
A Division of
Astral International Pvt. Ltd.
– ISO 9001:2008 Certified Company –
4760-61/23, Ansari Road, Darya Ganj,
New Delhi - 110 002
Phone: 23245578, 23244987
Fax: (011) 23260116
e-mail : dayabooks@vsnl.com
website : www.dayabooks.com

Laser Typesetting : **Classic Computer Services**
Delhi - 110 035

Printed at : **Chawla Offset Printers**
Delhi - 110 052

PRINTED IN INDIA

Preface

Neighther the strongest, nor the wisest survive, but he who survive on the earth who is responsive to the change. The eternal truth is the secret of success of any species on earth. Since inception, agriculture has gone through several transformations due to climatic, geographic, social, cultural, anthropological and several other reasons. In the last twenty years or so, global agriculture is going through rapid changes due to frequent changes in several factors. This put the agriculturists globally before tremendous challenges. There is a huge pressure to feed seven billion people and at the same time conserve and optimum utilization natural resources and sustain the agricultural production.

These also put forth a demand of several changes in the system of agricultural education, agricultural research and above all the extension activities. The knowledge, information, methods, process, practices etc., which proved successful in 60's and 70's are presently mostly wrong concepts on issue of sustainability. Due to these reasons several changes already occurred in the agricultural education, research and extension activities.

This book is primarily concerned about these changes in agricultural scenario in general and agricultural education, research and extension activities in particular. It provides basic knowledge and information about print and electronic media and its relation with extension education, different media credibility issues of farm information, different techniques of media research, study of its content etc. which will be helpful for the undergraduate students of Agricultural Sciences, Horticultural Sciences, Veterinary Sciences, Fisheries Sciences, Home Science etc. It will also provide an insight to the post-graduate and doctoral students of Agricultural Extension, Veterinary Extension, Fisheries Extension, Home Science Extension etc. and also for the students who are pursuing degree in media management studies, journalism and mass communication etc. this book also provide assistance to formulate research questions in the area of media credibility, content analysis, farm information, media management etc. It will also help the different media houses and media professionals of print and electronic media to work on the content of farm information, its credibility and can understand the importance of media in extension activities.

Thus, this book will fulfill the demand for the paradigm shift in the spectrum of agricultural education, research and extension activities and serve the purpose of the students, academicians, researchers, extension professionals, media professionals of varied disciplines.

B. Lahiri

S.D. Mukhopadhyay

M.M. Adhikary

Contents

List of Figures

List of Tables

Chapter 1

An Overview

The dawn of every civilization started with agriculture. Since time immemorial, every civilization has been acquiring knowledge and developing different techniques in the field of agriculture to feed its people. Though in present days, the strength of the economy of a nation is often presented by the minimum percentage of contribution of agriculture to the Gross Domestic Product of the country, but still its indispensable importance can not be ignored by any country. In the era of globalization, in India, a shifting trend of importance towards the trade and rapid industrialization rather than agriculture is quite conspicuous. This is mainly due to achievement of self-sufficiency in food grain production in the country. Over the last sixty years, India has achieved a remarkable progress in the field of agriculture from famine and chronic hunger to possessing more than thirty million tonnes of buffer stock of food grain. But there is every reason to think about that in an effort to be enlightened by the illumination of industrial development, rapid urbanization and eventually shifting focus from agriculture may lead to a disaster. This is because, though India has achieved self-sufficiency in food grain production but during last five years, the total food grain production has been fluctuating due to erratic rainfall and many other reasons. Though India has achieved to restrain her population growth rate but still we are adding every year a number, which is equal to the population of the Continent of Australia. Hence, it is not too far that we may have to face the shortage of food grain again if the present trend exists.

In this circumstance, we have not only to sustain our agricultural production but also to increase it for sake of feeding our future population. So there is no room for complacency of being self-sufficient in food grain production and agricultural development. But there is very limited scope of horizontal growth of agriculture in the country. Because very little arable land is left over which can be cultivated in India. At the same time, rapid industrialization and urbanization is encroaching the agricultural land. Situation has compelled us to think about vertical growth of agriculture. But here also, the picture is not so rosy. Due to intensive cultivation, injudicious application of inorganic fertilizers and agrochemicals and many other reasons, our soil fertility status is depleting day by day. At the same time excessive pressure on environment and natural resources and its detrimental effects on biosphere make situation more critical. Although, Indian agriculture has to grow vertically, which calls more adoption of the latest and appropriate farm technologies at field level, as there exists a significant adoption gap at the field level. So a well planned and efficiently executed extension policy can address the problem and can play a pivotal role in the sustainable development of agriculture in India. In this regard, farm information communication has greater role to play and towards that, mass media can be of immense importance.

Since independence, several steps have been taken in the domain of farm information communication, which obviously paid dividends, witnessed by remarkable development in the field of agriculture. But most of these attempts were traditional in approach. Generally, in India, farm information were tried to disseminate to farm people through organizational and administrative set up. In the early stage of development after independence in India, specific target area and target group were addressed and farm information were tried to disseminate to that specific area and specific group of people. In the process a large portion of the population remained untouched. Some steps were also taken for area expansion to cover the larger section of the rural society. But these steps never yielded fruitfully. Probably it is one of the reasons of confinement of fruits of Green Revolution to a limited part and limited section of rural society. In 1974, Training and Visits System was introduced in India. But here again, farm information were disseminated to only contact farmers who had the responsibility to disseminate those information to the larger group of the farmer. In this process larger area was covered but again larger group of people remain deprived due to poor agricultural administrative set up and limited staff strength. Eventually our farmers have not adopted a large percentage of our research findings and even that information did not reach to them. So in a vast country likes India with huge and diversified population, it is always very tough task to reach the larger mass of the rural society at a time. To reach this huge population, mass media have the tremendous potentiality, which has not been exploited till now in the field of farm information communication.

The Tenth Plan (2002-2007) Document is recognizing the problems of the extension sector and states:

"The Agriculture extension machinery and information support in most states seems to have become outmoded. The staff created under the World Bank assisted Training and Visits (T& V) programme do not have much mobility. The need to revamp the extension services in the country by using print and electronic media and information technology along with the involvement of the private sector, especially the input agencies and traders, are now one of the main sources of information for the farmers. Radio, Television and the print media have become powerful means of education and technology dissemination."

The public extension programme being state subject, are struggling due to financial constraints and therefore have not been able to effectively carry out extension activities. A situation has thus arisen where the need for providing quality extension service to the farmers is the greatest while these services have shrunken to almost non-existent level. The private extension services, though effective, have not been able to cover the complete spectrum of agricultural activities. They have been predominately limited to certain high valued crops in resource-endowed areas confined to progressive farmers. The small and economically weak farmers are not yet able to pay for the services offered by private entrepreneurs. At the same time, in a country like India, where two-third of the population depend on agriculture and one-fourth of countries GDP is contributed by agriculture, it is very upsetting to note that the coverage of agriculture in media including print and electronic which is even less than two per cent. This is probably farm information is not lucrative to the media organizer. As non-glamorous information it is also non-profitable to them. Driven by the facts of poor extension services and apathy of media for agriculture, the Ministry of Agriculture, Government of India has taken a bold initiative to sponsor scheme on Mass Media Support to Agriculture Extension and which is already under implementation from 2003-2004. Its primary objective is to use television and radio with their massive penetration as a vehicle that could be exploited for the purpose of extension. In this regard, two different mass media, newspapers and radio have certain edge over other mass media in terms of agricultural extension due to wide coverage and higher degree accessibility by our resource poor farming community.

Newspapers and radio has long history in India. Kolkata, as one of the most important city under British India, most of pioneering work in the development of newspaper and radio started in Kolkata. The newspaper movement in India started with the *Hickey's Bengal Gazette or the Calcutta General Advertiser* by James Augustus Hickey in 1780. Since then lot of progress has been achieved in the development of newspapers in West Bengal and India as well. Focus of this media has changed time to time since its inception. Newspapers has played important role in the development of agriculture in the country. Newspaper demands a moderate level of literacy but not high level of education. In India literacy rate has been rising in a steady rate. So newspapers have a very good potentiality in terms of dissemination of farm information. In an agriculturally advanced state like West Bengal, newspapers have lot of contribution in the agricultural extension activities. In West Bengal, several daily newspapers have been publishing. *Aajkal, Ananda Bazar Patrika, Bartaman* and *Sambad Pratidin* are the most widely circulated newspapers among them. Same way, radio broadcasting started in India in 1927 with two privately-owned transmitters at Mumbai and Calcutta (presently known as Kolkata) which were taken over by the Government in 1930. It also came to be known as "Akashwani" from 1957. But still it is popular as *All India Radio* in India. AIR has the privilege to cover 99.37 per cent of India's populace. *All India Radio, Kolkata* also covers almost every part of West Bengal and huge Bengali spoken people in West Bengal and in the country as well. *All India Radio, Kolkata* broadcasts three farm programmes *viz., Chasi Bhaider Bolchhi, Uno Jamir Duno Phasal* and *Krishi Kathar Aasar* everyday. Being cost effective and wide area coverage has some relative advantages over other media. At the same time radio may even prove more potential media for the farm information dissemination.

In the present context of increasing importance of mass media for communication of farm information, what is expected is that an appropriate content, process, structure and system have to be pragmatically developed. For that purpose a detailed study in relation to farm information communicated through mass media with special reference to the content wise nature and extent of farm information through mass media, like; basis of ascertaining content to be communicated, consideration of readers' needs for the farm information, extent of credibility of farm information from scientific and users' point of view, different factors there off and so on, are of immense importance. The four widely circulated Bengali newspapers of West Bengal, India; *Aajkal, Ananda Bazar Patrika, Bartaman* and *Sambad Pratidin* and three farm programmes of *All India Radio, Kolkata* have been disseminating a considerable number of farm information for the farming community. But, at the same time, the farmers' perception regarding this farm information should also be addressed to get a more vivid picture regarding the matter.

Regarding the topic some basic assumptions of farm information communicated through above mentioned newspapers and radio programme will help to understand the subject matter in more conspicuous way:

- ☆ The stated mass media, newspapers and radio communicate farm information.
- ☆ All the newspapers and radio programme are not similar in terms of farm information communication.
- ☆ Media have their own preferences for selection of different major categories and sub-categories related to farm and farm business for farm information communication.
- ☆ Differential importance is being given by different newspapers and different radio programme towards farm information communication.
- ☆ Farm information seeking behaviour is differential in nature.
- ☆ Mass media like newspapers and radio have differential credibility from technical as well as farmers' point of view.

- Farmers' socio-economic and socio-personal variables have relationship with mass media credibility.

On the basis of the above logic and assumptions, some objectives have been framed:

- To study the comparative orientation and importance given to farm information by the stated newspapers and radio programme.
- To study the farm information seeking behaviour of the farmers.
- To study the credibility of farm information communicated through newspapers and radio from technical point of view.
- To study the credibility of farm information communicated through Bengali newspapers and radio as perceived by the farmers.
- To enumerate the relationship of different socio-economic and socio-personal variables with the credibility of farm information.
- To study the differential perception of credibility of farm information communicated through stated newspapers and radio programme from technical and farmers' point of view and;
- To suggest some policy measures.

Chapter 2
Theoretical Backdrop

Any successful scientific study always needs a comprehensive review of literature. Its indispensable relevancy lies in the fact that it helps:

1. To provide a basis for the development of theoretical frame work,
2. To provide an insight in to the methods and procedures,
3. To suggest an operational definition, and
4. To provide a basis for the interpretation of findings.

Dissemination of Farm Information through Mass Media

Dissemination of farm information through mass media is a very important aspect of extension activities in any country. The importance of mass media in farm information communication was acknowledged by various researchers in the past. Rogers and Beal (1958), opined that mass media were generally most important in creating awareness of new ideas. Among the different mass media newspapers and radio are two very important media in terms of farm information communication in a resource poor developing country. Chiders and Vajrathan (1973) indicated that the mass media, especially the printed media, are extremely important in stimulating a sense of environment. Dey (1976) wrote, "in comparison with the other media, Newspaper is the oldest mass medium". Besides, Radio and Television have created significant influence on the society. In some field of communication the Radio and Television have crossed the scope of Newspaper.

In India, where there exist a significant gap in knowledge about latest farm information and subsequent adoption gap, there mass media have a very important role to play. Since long back farm information are being communicated through newspapers and radio.

Newspaper: The Indian Scenario

William Bolts, an ex-employee of the British East India Company attempted to start publishing the first in India in 1776. Bolts had to beat a retreat under the disapproving gaze of the Court of Directors of the Company. The *Hickey's Bengal Gazette or the Calcutta General Advertiser* was started by James Augustus Hickey in 1780. The *Gazette*, a two-sheet newspaper, specialized in writing on the private lives of the Sahibs of the Company. As the cultural and literal capital of India, Kolkata was the breeding ground of newspaper movement in India. Several newspapers, the local vernacular dailies in particular played a significant role in the Indian National Movement. British Government even compelled to enact Vernacular Press Act to restrict the vernacular newspapers to agitate the people against the colonial British Government. So newspapers in India and also in West Bengal have a rich heritage. In present days, Press Trust of India (PTI) and United News of India (UNI) are the two primary Indian news agencies. The former was formed after it took over the operations of the Associated Press of India and the Indian operations of Reuters soon after independence on August 27, 1947. PTI is a non-profit cooperative of the Indian newspapers.

In West Bengal, several daily newspapers have been publishing. Among those, four widely circulated newspapers, *Aajkal, Ananda Bazar Patrika, Bartaman and Sambad Pratidin* have been taken for the study.

Information about All India Radio

All India Radio (AIR for short), officially known as *Akashwani* is the radio broadcaster of India and a division of *Prasar Bharati* (Broadcasting Corporation of India), an autonomous corporation of the Ministry of Information and Broadcasting, Government of India. *All India Radio* is one of the largest radio networks in the world. Its headquarters is at Akashwani Bhavan in New Delhi, on the Parliament Street next to the Indian parliament. Broadcasting house is an old building next to Akashwani Bhavan. The News Service Division of All India Radio under the Director General Now functions from this building. Built during the British rule, it is a very popular location and easily recognized building in New Delhi.

History

Broadcasting started in India in 1927 with two privately owned transmitters at Mumbai and Calcutta, which were taken over by the Government in 1930. These were operating under the name "Indian Broadcasting Service" until 1936 when it was given the present name "All India Radio (AIR)". It also came to be known as "Akashwani" from 1957. The Five Year Plans have given new impetus to the growth of broadcasting resulting in a phenomenal expansion from 6 stations at the dawn of Independence to around 200 stations at the close of the millennium.

Coverage

AIR covers 99.37 per cent of India's populace, the largest democracy in the world with over one billion population. AIR maintains approximately 200 broadcasting centres around the country and transmits in 24 different languages. In spite of recent penetration by other media such as Cable TV, AIR remains the most common means of gaining access to information and entertainment, as the radio receivers are relatively cheap and affordable.

Services

AIR has many different services each catering to different regions/languages across India. One of the most famous services of the AIR is the *Vividh Bharati Seva* (roughly translating to "Multi-Indian service"). This service is the most commercial of all and is popular in all cities of India. This service offers a wide range of programmes including news, film music, comedy shows, etc. The *Yuv-vani*(The Voice of Youth) service of AIR provides an enriching and novel radio-experience by encouraging youth participation and experimenting with varied script ideas. With shows like *"Mehfil"* (The Party), "In the groove" and "The Roving Microphone" which have been around for more than three decades, *Yuv-vani* still holds a firm ground of its own. All

India Radio, after launching the news-on-phone service on 25th February 1998 from New Delhi, is now running the service from Chennai, Mumbai, Hyderabad and Patna also. The service is accessible through STD, ISD and local telephone calls. The service is going to be started from 10 more cities – Ahmedabad, Banglore, Guwahati, Imphal, Jaipur, Kolkata, Lucknow, Raipur, Simla and Thiruvanthapuram shortly.

The main projects that describe the growth of educational radio are:

School Broadcast Project

This project was commissioned in 1937 and the target group was School students. This programme started from Delhi, Calcutta, Madras and Bombay. In the beginning the school programme were not strictly governed by the curriculum. With the passages of time and acquisitions of more experience, the AIR tried to make its radio broadcasts more curriculum oriented, but in absence of common syllabi and timetables in schools, even within the same state, it could not succeed in its aim.

Adult Education and Community Development Project (Radio Forum)

Commenced in 1956, the Villagers of 144 villages in the vicinity of Poona (in Maharastra state), were the main beneficiaries of this project. This was agriculture-based project, which was originally designed and tried out in Canada. With the help of UNESCO, it was tried in 144 villages of Poona and was named as ' Radio forums Project' (defined as a listening cum-discussion-cum-action group). The members of the forum could listen thirty-minute radio programme on some agricultural or community - development programme, then discuss and decide regarding it's adoption in their own village. This project was a great success. Many action programmes were planned and put into practice.

Farm and Home Broadcast Project

This project was commenced in 1966 and again targeted at farmers and villagers. These broadcasts were designed to provide information and advice on agricultural and allied topics. The aim was to educate the farmers and provide them assistance in adopting innovative practices in their fields as per the local relevance. The experts also conducted occasional farm radio schools, which proved to be very effective.

University Broadcast Project

This project for University students was initiated in 1965, with an aim to expand higher education as widely as possible among the different strata of society. The Programme consisted of two types- 'General' and 'enrichment'. The general programmes included topics of public interest and enrichment programmes supported correspondence education offered by Universities in their respective jurisdictions. School of Correspondence Studies, University of Delhi and the Central Institute of English and Foreign Languages, Hyderabad is well known for preparation and broadcast of their programmes through AIR.

Language Learning Programme

The project, popularly known as 'Radio Pilot Project' was started in 1979-80 jointly by AIR and Department of

Education, Government of Rajasthan, with an aim to teach Hindi to School going children as first language in 500 primary schools of Jaipur and Ajmer districts on experimental basis.

IGNOU-AIR Broadcast

In collaboration with IGNOU, AIR stations of Mumbai, Hyderabad and Shillong started radio broadcasts of IGNOU Programmes from January 1992. Main target group of this project were students of Open/Conventional Universities.

IGNOU-AIR Interactive Radio Counselling (IRC)

Started in 1998 for students of Open/Conventional Universities, this project is also very successful. In order to bridge the gap between Institutions and learners by instantly responding to their queries and also to provide Academic Counselling in subject area, IGNOU in collaboration with AIR Bhopal started this project in May 1998 as an experimental programme for one year (Sharma, 2002a). With the success of the experiment, it was extended to 8 other AIR stations (Lucknow, Patna, Jaipur, Shimla, Rohtak, Jalandhar, Delhi and Jammu). Presently Interactive Radio counseling is being provided on every Sunday for one hour (4: 00 PM - 5: 00 PM) from 186 radio stations of All India Radio. Toll-free telephone facility is available from 80 cities (effective from February 2001) enabling the learners to interact with experts and seek clarification, without paying for their telephone calls.

Gyan-Vani (Educational FM Radio Channel of India)

This project is recently launched (in year 2001) and again the target group is students of Open/Conventional

Universities. Gyan-Vani (Gyan = Knowledge, Vani = aerial broadcasting) is Educational FM Radio Channel of India, a unique decentralized concept of extending mass media for education and empowerment, suited to the educational needs of the local community (Sharma, 2002b). It is operating presently through Allahabad, Bangalore and, Coimbatore FM stations of India on test transmission mode. The network is slotted to expand to a total of 40 stations by June-2002. Gyan-Vani, in addition to giving the hardcore education also deals with awareness programmes including the ones for Panchayati Raj Functionaries, Women Empowerment, Consumer Rights, Human Rights, the Rights of the Child, Health Education, Science Education, Continuing Education, Extension Education, Vocational Education, Teacher Education, Non-formal Education, Adult Education, Education for the handicapped, Education for the down trodden, education for the tribals and so on. Gyan-Vani is available through commercial FM radio set.

Radio-Vision (Multimedia through Digital Radio)

Pioneered by the BBC, the technique of radio-vision allows the subject matter to be presented through two channels, the audio and the visual. The visuals are presented in the form of still filmstrips, charts, slides, models, etc, while the explanation is given through recorded narration. Educational institutions use this as a substitute for educational television. Radio-vision has its own advantages:

- It is economical
- It can cater to different categories of learners
- It is easy to produce such programmes at the institutional level or at the learning centers

- ☆ It provides visual support to the concept that is taught.

(*Source*: Vyas, 2002)

Content Analysis

Content analysis is a very useful technique, often applied by different researchers in the field of Journalism and Mass Communication and other various field of research activities.

The term "Content analysis" is about 60 years old. Webster's dictionary of the English Language included the term in its 1961 edition, defining it as analysis of the manifest and latent content of a body of communicated material (as a book or film) through classification, tabulation, and evaluation of its key symbols and themes in order to ascertain its meaning and probable effect.

According to Leites and Pool (1942), "Content analysis" requires inferences be derived strictly from counts of frequency place a number of standard methods on the borderline of acceptability".

According to James (1949), "Content analysis" may be defined as "referring to any technique (a) for the classification of the sign-vehicles, (b) which relies solely upon the judgements (which theoretically, may range from perceptual discriminations to sheer guesses) of an analyst or group of analysts as to which sign-vehicles fall into which categories, (c) on the basis of explicitly formulated rules (d) provided that the analyst's judgements are regarded as the reports of a scientific observer. Some definitions of Content analysis, equate it with numerical".

Cartwright (1953), proposed to use the terms "content analysis" and "coding' inter-changeably to refer to the objective, systematic, and quantitative description of any symbolic behaviour.

Barcus (1959), used the term "Content analysis" to mean the scientific analysis of communications messages. The method is broadly speaking the "Scientific method", and while being catholic in nature, it requires that the analysis be rigorous and systematic.

Osgood (1959), defined "Content analysis" as "a procedure whereby one makes inferences about sources and receivers from evidence in the messages they exchange".

Stone *et al.* (1966), stated "Content analysis is any technique for making inferences by systematically and objectively identifying specified characteristics within text."

Weber (1990) defined "Content analysis is a research method that uses a set of procedures to make valid inferences from the text. These inferences are about sender(s) of the message, that message itself, or the audience of the message. The rules of this inferential process vary with the theoretical and substantive interests of the investigator".

Krippendorff (2004) defined "Content analysis is a research technique for making replicable and valid inferences from text (or other meaningful matter) to the context of their use.

Krippendorff (2004) opined that contemporary content analysis has three distinguishing characteristics.

First, "Content analysis" is an empirically grounded method, exploratory in process, and predictive or inferential in intent.

Second, contemporary content analysis transcends traditional notion of symbols, contents and intents.

Third, contemporary content analysis has been forced to develop a methodology of its own.

Krippendorff (2004) categorized Content analysis used for different purpose are:

1. Extrapolation
2. Standards
3. Indices and symptoms
4. Linguistic representations
5. Conversation
6. Institutional Process.

Krippendorff (2004) suggested that the different components of content analysis are:

1. Unitizing
2. Sampling
3. Recording/Coding
4. Reducing data to manageable representation
5. Abductively inferring contextual phenomena
6. Narrating the answer to the research question.

Credibility Study

Researchers have utilized a variety of measurements and statistical procedures in their quest to understand media credibility. Bivariate and multivariate approaches have been used, including regression analysis (Mulder, 1980; Mulder, 1981), and factor analysis. While many have used traditional data-collection methods such as telephone surveys and laboratory and field experiments. Infante (1980) used three dimensions to measure source credibility. These were trustworthiness, expertise and dynamism. Trustworthiness was operationalized as honest-dishonest, trustworthy-untrustworthy and sincere-insincere. For expertise, he used skilled-unskilled, qualified-unqualified and informed-uninformed. For dynamism, he used bold-timid, active-passive, and aggressive-meek.

Gaziano and McGrath (1986), observed that media credibility is comprised of fairness, (un)bias, telling the whole story, accuracy, respect for privacy, watching out after people's interest, concern for community well-being, separation of fact and opinion, trustworthiness, concern for public interest, factuality and reporter training level. They identified twelve dimensions of newspaper and television news credibility which are fairness, bias, completeness, accuracy, respect for privacy, watch for people's interest, concern for community, separation of fact and opinion, trust, concern for public interest, factual and level of training. Furthermore, Gaziano's (1987) analysis of four major credibility studies found twelve operationalizations of credibility. These included believability; accuracy, completeness, and covering up facts; trustworthiness and reliability; being unbiased, balance of coverage, fairness, objectivity; other characteristics of press performance, such as invasion of privacy, covering up of stories; overall evaluation of how well media perform; confidence in media institutions, comparisons of media with other institutions; independence of media from special interests, other organizations, institutions; power/influence of media in community or society; relationship of news media to government; honesty and ethical standards; and professionalism, training of people in the media. It can also be noted that these measures had also been used in studies by Hovland and Weiss (1951), Meyer (1988), and others.

Rimmer and Weaver (1987) reported a Cronbach alpha of 0.90 for the Gaziano and McGrath scale for both newspapers and television. Trustworthiness, fairness, bias, completeness, respect for privacy, representation of individual interests, accuracy, concern for community well-being, separation of facts and opinion, concern for public interest, factual foundations of information published and qualifications of reporters were used among the credibility measures by Rimmer and Weaver (1987).

Meyer (1988) criticized the Gaziano and McGrath scale as lacking face validity and theoretical grounding. He developed a five-item news credibility scale. The items are fair, unbiased, tells the whole story, accurate, and can be trusted and yielded a Cronbach alpha of 0.83. On the basis of index for newspaper believability was developed, which comprised of five variables. These include fairness, bias, completeness, accuracy and trustworthiness.

Wanta and Hu (1994), used believability and affiliation indices to evaluate media credibility. The believability index was built around media manipulation of public opinion, getting facts straight, dealing fairly with all sides of an issue and separation of fact from opinion. Affiliation was measured with concern for community well being, watching out for reader interests, and concern for public welfare.

Ognianova (1998) utilized nine semantic differential items to measure online news story credibility. They were factual/opinionated, fair/unfair, accurate/inaccurate, untrustworthy/trustworthy, balanced/unbalanced, biased/unbiased, reliable/unreliable, thorough/not thorough, and informative/not informative.

Johnson and Kaye (1998, 2000), employed believability, fairness, accuracy and depth of information in their study. Sensationalism was one of six dimensions used by Sundar (1996). He also used accuracy, believability, bias, fairness and objectivity. Sundar (1999) developed a credibility scale applicable to both newspapers and online newspapers. He found a striking similarity between the factor structures underlying receiver's perceptions of print and online newspapers. He claimed that this similarity made it possible to use the same scales for different media. Schweiger (1998) pointed out that credibility becomes an important heuristic for content selection at a time of information overload. Credibility may also influence the journalistic and commercial success of a medium (Schweiger, 2000).

Flanagin and Metzger (2000) used single-item measures in studying the credibility of Internet information. They operationalized credibility as a multi-dimensional concept built from five traditional components found in the literature: believability, accuracy, trustworthiness, bias, and completeness.

Definitions of Different Terms Used

Farm Information

Before operationalizing 'Farm Information', definition of the words 'Farm' and 'Information' are separately given below:

Farm

A farm is the basic unit in agriculture. It is a section of land devoted to the production and management of food.

It may be an enterprise owned and operated by a single individual, family, or community, or it may be owned by a corporation or company. A farm can be a holding of any size from a fraction of a hectare to several thousand hectares. In England, there is a vague point when a large farm ceases to be referred to as a farm and becomes an estate.

(*Source*: *www.en.wikipedia.org/wiki/Farm*)

- ☆ Any activity conducted solely or primarily for the production of one or more agricultural products or commodities, including timber, for sale and home use, and customarily producing such products or commodities in sufficient quantity to be capable of contributing materially to the operator's support.

(*Source*: *www.fhwa.dot.gov/realestate/rights/defin.html*)

Information

A message received and understood.

(*Source*: *www.wordnet.princeton.edu/perl/webwn*)

- ☆ Information is a term with many meanings depending on context, but is as a rule closely related to such concepts as meaning, knowledge, instruction, communication, representation, and mental stimulus.

(*Source*: *www.en.wikipedia.org/wiki/Information*)

- ☆ Data that have been processed and presented in a form suitable for human interpretation, often with the purpose of revealing trends or patterns.

(*Source*: *www.gtscompanies.com/glosscomp.html*)

- ☆ Information is the result of processing, manipulating and organizing data in a way that adds to the knowledge of the person receiving it.

(*Source*: *www.orafaq.com/glossary/faqglosi.htm*)

News

There are many definitions of 'News'. Some of which are given under:

- ☆ New information about specific and timely events; "they awaited news of the outcome"

(*Source*: *www.wordnet.princeton.edu/perl/webwn*)

- ☆ 'News' is the reporting of current events usually by local, regional or mass media in the form of newspapers, television and radio programs, or sites on the World Wide Web. News reporting is a type of journalism, typically written or broadcast in news style. Most news is investigated and presented by journalists (or reporters) and often distributed via news agencies. If the content of news is significant enough, it eventually becomes history.

(*Source*: *www.en.wikipedia.org/wiki/News*)

Views

'Views' are basically the interpretation of some event or phenomena by an individual or group. In that sense 'Views' may be considered as opinion. There are many definitions of opinion; some of which are described below:

- ☆ **Opinion:** a personal belief or judgment that is not founded on proof or certainty; "my opinion differs from yours"; "what are your thoughts on Haiti?"

- **Opinion:** a message expressing a belief about something; the expression of a belief that is held with confidence but not substantiated by positive knowledge or proof; "his opinions appeared frequently on the editorial page".

(*Source*: *www.wordnet.princeton.edu/perl/webwn*)

- An external representation of an internal processing model which is used to interact with the internal model.

(*Source*: *www.it.bton.ac.uk/staff/je/adacraft/glossary.htm*)

Advertisement

There are several definitions of advertisement suitable for different media. Some of them are:

- A public promotion of some product or service.

(*Source*: *www.wordnet.princeton.edu/perl/webwn*)

- Generally speaking, advertising is the paid promotion of goods, services, companies and ideas by an identified sponsor. Marketers see advertising as part of an overall promotional strategy. Other components of the promotional mix include publicity, public relations, personal selling, and sales promotion.

(*Source*: *www.en.wikipedia.org/wiki/Advertisement*)

- Item of publicity to promote a product or service in newspapers, magazines, on TV etc.

(*Source*: *www.learn-english-today.com/business-english/advertising-vocabulary.html*)

- A message printed in the newspaper in space paid for by the advertiser.

(*Source*: *www.angelfire.com/trek/puknews/refsnp.html*)

Deliberation

Some of the definitions of deliberation are:

- ☆ (Usually plural) discussion of all sides of a question.

  (*Source*: *www.wordnet.princeton.edu/perl/webwn*)

- ☆ The consideration, discussion and decision making process, as by a jury.

  (*Source*: *www.attorneykennugent.com/library/d.html*)

Interview

There are several kinds of interviews generally used in our society. Definition of interview is:

- ☆ An interview is a conversation between two or more people where questions are asked to obtain information about the interviewee. Interviews can be divided into two rough types, interviews of assessment and interviews for information.

  (*Source*: *www.en.wikipedia.org/wiki/Interview*)

Radio Play

Radio play or radio drama is popular programme in every parts of the world. Definition of radio play or radio drama is:

- ☆ Radio drama, which had its greatest popularity in the U.S. and in most other countries before the widespread access to television programming, depends on dialogue, music and sound effects to help the listener imagine the story in her or his "mind's eye"–in this sense; it resembles reading in some ways more than audio/visual drama. In the television era, some audio drama has been produced and marketed that owes much to radio drama, without ever being broadcast on radio.

(*Source*: *www.en.wikipedia.org/wiki/Radio_play*)

Panel Discussion

There are several definitions of panel discussion. Some of which are as follows:

- ✰ A structured conversation on a given topic among several people in front of an audience.

  (*Source*: *www.highered.mcgraw-hill.com/sites/007256296x/student_view0/glossary.html*)

- ✰ Instructional Technique using a group of people chosen to discuss a topic in the presence of an audience.

  (*Source*: *www.accenter.com/aboutthecenter/glossary.asp*)

- ✰ Discussion of a subject of public interest by a group of persons forming a panel usually before an audience.

  (*Source*: *www.wordnet.princeton.edu/perl/webwn*)

Agricultural Policy

There are several kinds of policies which are formulated or enforced by different agencies in our society. The definition of agricultural policy is:

- ✰ Agricultural policy is a loose term that describes a government's set of laws relating to domestic agriculture and imports of foreign agricultural products. Governments usually implement agricultural policies with the goal of achieving a specific outcome in the domestic agricultural product markets. The specific outcomes can range from guaranteed supply level, price stability, product quality, product selection, land use or employment.

(*Source*: *www.en.wikipedia.org/wiki/Agricultural_policy*)

Agricultural Pollution

Pollution is one of the greatest menaces of the present world. Agricultural pollution has significant contribution in global pollution. Definition of the agricultural pollution is:

- ☆ Farming wastes, including run off and leaching of pesticides and fertilizers; erosion and dust from plowing; improper disposal of animal manure and carcasses; crop residues, and debris.

(*Source*: *www.e11th- hour.org/resources/backgrounders/ environmental.glossary.a.html*)

Agricultural Crisis

Different kinds of crisis generally occur in our human society. Agricultural crisis is also one of them. Definitions of agricultural crisis are as follows:

- ☆ An unstable situation of extreme danger or difficulty; "they went bankrupt during the economic crisis"
- ☆ A crucial stage or turning point in the course of something; "after the crisis the patient either dies or gets better".

(*Source*: *www.wordnet.princeton.edu/perl/webwn*)

- ☆ «A time when a problem or situation is very bad or dangerous» [Longman] «a decisive moment; a time of danger or great difficulty» [Oxford]

(*Source*: *www.terena.nl/tech/task-forces/tf-csirt/iodef/docs/ i-taxonomy_terms.html*)

Agricultural Marketing

Marketing is an important component in any production process in present day's world. Agricultural marketing also plays vital role in the agriculture. Some of the definitions of agricultural marketing are as follows:

- Means to make a communication about a product or service a purpose of which is to encourage recipients of the communication to purchase or use the product or service.

  (*Source*: *www.preemptinc.com/definitions.html*)

- The planning and implementation of a strategy for the sale, distribution, and servicing of a product or service.

  (*Source*: *www.mvp.cfee.org/en/glossary.html*)

- Traditionally, Marketing has been a term applied to the craft of linking the producers (or potential producers) of a product or service with customers, both existing and potential. This general definition fails to provide any direction to someone hoping to market their products or services effectively. With that, a more modern explanation of what marketing is–and what makes it effective–is a definition coined by marketing researcher and author Brian Norris.

  (*Source*: *www.en.wikipedia.org/wiki/Marketing*)

Natural Calamities

There are several kinds of natural calamities which have happened on earth. Definitions of natural calamities are as under:

- An event resulting in great loss and misfortune; "the whole city was affected by the irremediable calamity"; "the earthquake was a disaster".

  (*Source*: *www.wordnet.princeton.edu/perl/webwn*)

- A disaster is a natural or man-made event that negatively affects life, property, livelihood or industry often resulting in permanent changes to human societies, ecosystems and environment.

  (*Source*: *www.en.wikipedia.org/wiki/Calamity*)

Social Conflict

Several kinds of social conflicts occur in the society and also in rural settings. There are several definitions of social conflict:

- A state of opposition between persons or ideas or interests; "his conflict of interest made him ineligible for the post"; "a conflict of loyalties".
- Dispute: a disagreement or argument about something important; "he had a dispute with his wife"; "there were irreconcilable differences"; "the familiar conflict between Republicans and Democrats".
- Social conflict is a conflict or confrontation of social powers.

  (*Source*: *www.wordnet.princeton.edu/perl/webwn*)

Social Forestry

Social forestry is a very important dimension of forestry which helps the society. Some definitions of social forestry are as follows:

- ☆ The science, art and practice of managing and using trees, forests and their associated resources for human benefit.

(*Source*: *www.calforests.org/glossary.html*)

- ☆ The science, art and practice of managing and using for human benefit the natural resources that occur on and in association with forest lands.

(*Source*: *www.borealforest.org/nwgloss4.htm*)

Success Story

There are several example of success stories can be found in our society. Definition of success story is as follows:

- ☆ An example of a successful practice – the company has achieved documented successful results. (Stage: This term is used to distinguish those companies that have succeeded with a practice from those that are experimenting (cf. "Experiment") with one or a documented failure (cf. "Failed Experiment").

(*Source*: *www.ccs.mit.edu/21c/iokey.html*)

Technology Transfer

Technology transfer generally happens in various dimensions in our present day's world. Following definitions will give an idea about technology transfer:

- ☆ Activities that lead to the adoption of a new technique or product by users and involve dissemination, demonstration, training and other activities that leads to eventual innovation.

(*Source*: *www.oahutrans2k.com/info/glossary/T.htm*)

- ☆ Technology transfer is understood to be the passing of theoretical and practical skills and know-how from the owner of a technology to outside users or beneficiaries of technology. Technology transfer is in comparison to technology cooperation an isolated and time-limited transaction. Technology transfer is not simply about the supply and shipment of hardware across international borders. It is about the complex process of sharing knowledge and adapting technology to meet local conditions. It strengthens human and technological capacity in developing countries.

(*Source*: *www.lineadecreditoambiental.org/html/glossary.html*)

Rural Development

Development of the rural society is an important aspect of the development policy in any country. Definition of rural area is as follows:

- ☆ Projects located in rural counties or in a metropolitan county with a resident population less than 20,000 and the SBA has determined such area to be rural.

(*Source*: *www.preferredlendingpartners.org/programterms.htm*)

Weather Forecast

Weather forecast is generally communicated by different mass media and other agencies to the people for various needs. Some of the definitions of weather forecast are as follows:

- ☆ An assessment of the future state of the atmosphere with respect to precipitation, clouds, winds, and temperature. Such assessments are usually made by government or private meteorologists, often using numerical simulations. Such simulations are the result of representing the atmosphere mathematically as a fluid in motion. See also numerical weather prediction.

(*Source*: *www.amsglossary.allenpress.com/glossary/browse*)

☆ A prediction of weather conditions expected at a place, within an area, or along a route for a specified time or during a specified period.

(*Source*: *www.globalsecurity.org/wmd/library/policy/army/fm/3-6/3-6gl.htm*)

Agricultural Machinery

Agricultural machinery is the indispensable part of the agriculture. Definition of agricultural machinery is as follows:

☆ Agricultural machinery is one of the most revolutionary and impactful applications of modern technology. Given the truly elemental human need for food, agriculture has been an essential human activity almost from the beginning, and it has often driven the development of technology and machines. Over the last 100 years, advances in farm equipment have dramatically changed the way people are employed and produce their food worldwide.

(*Source*: *www.en.wikipedia.org/wiki/Agricultural_machinery*)

Fertilizer

In modern day's agriculture, fertilizer is one of the most important inputs. Some of the definitions of fertilizer are as follows:

- Organic or inorganic plant foods which may be either liquid or granular, used to amend the soil in order to improve the quality or quantity of plant growth.

  (*Source*: *www.thegardenhelper.com/dictionary.html*)

- A single or blended substance containing one or more recognized plant nutrient(s) which is used primarily for its plant nutrient content and which is designed for use or claimed to have value in promoting plant growth.

  (*Source*: *www.ebfarm.com/Organic/Glossary.aspx*)

- Any organic or inorganic material of natural or synthetic origin that is added to the soil to supply certain elements essential to the growth of plants.

  (*Source*: *www.ipmalmanac.com/glossary/index.asp*)

Fishery

Definition of fishery is as follows:

- A fishery (plural: fisheries) is an organized effort by humans to catch fish or other aquatic species, an activity known as fishing. Generally, a fishery exists for the purpose of providing human food, although other aims are possible (such as sport or recreational fishing), or obtaining ornamental fish or fish products such as fish oil. Industrial fisheries are fisheries where the catch is not intended for direct human consumption.

(*Source*: *www.en.wikipedia.org/wiki/Fishery*)

Seed

Seeds or seed materials is another important component of agriculture. It can be defined as:

- ☆ Botanically, a seed is the matured ovule without accessory parts. Colloquially, a seed is anything which may be sown; ie, seed potatoes (which are vegetative tubers); seed of corn, sunflower, etc.

  (*Source*: *www.fao.org/docrep/003/X3910E/X3910E22.htm*)

- ☆ The part of a plant that can make another plant. Usually a plant makes flowers, then once the flowers have been pollinated, they turn into fruit. The seeds are inside the fruit. Seeds can be spread by wind, water, or animals. Then, when they are in a new place with water and nutrients, can grow into a new plant. Sometimes an animal eats some fruit, and then poops the seeds out someplace, and a new plant grows there!

  (*Source*: *www.fcps.k12.va.us/StratfordLandingES/Ecology/mpages/glossary_qz.htm*)

Co-operative

Several kinds of co-operatives, can be found in our society. Some of the definitions of co-operative are as under:

- ☆ A member-owned business with membership opens to those who use its services. Democratically controlled and operated on a nonprofit basis, a cooperative returns any margins or profits to members on the basis of patronage.

(*Source*: *www.dairynet.com/kids/gloss.html*)

- ✰ A duly registered association of persons, with a common bond of interest, who have voluntarily joined together to achieve a lawful common social or economic end, making equitable contribution to the capital required and accepting a fair share of the risks and benefits of the undertaking in accordance with universally accepted cooperative principles.

(*Source*: *www.nscb.gov.ph/ru12/DEFINE/DEF-AGRI.HTM*)

Health and Sanitation

Before operationalizing the 'Health and Sanitation' the definitions of 'Health' and 'Sanitation' are described below separately:

Health

A state of complete physical, mental, and social well-being and not merely the absence of disease or infirmity.

(*Source*: *www.cdc.gov/nccdphp/drh/epi_gloss.htm*)

- ✰ A dynamic state of complete physical, mental, spiritual and social wellbeing and not merely the absence of disease or infirmity. (WHO'S New Proposed Definition. 101st Session of the WHO Executive Board, Geneva, January 1998. Resolution EB101.R2).

(*Source*: *www.mapp.naccho.org/mapp_glossary.asp*)

- ✰ Health is defined in the WHO constitution of 1948 as: 'A state of complete physical, social and mental well-being, and not merely the absence of disease or infirmity. Within the context of health promotion, health has been considered less as an abstract state and more as a means to an end, which can be expressed, in functional terms as a resource, which permits people to lead an individually, socially and economically productive life. Health is a resource for everyday life, not the object of living. It is a positive concept emphasizing social and personal resources as well as physical'.

(*Source*: *www.health.nsw.gov.au/public-health/health-promotion/abouthp/glossary.html*)

Sanitation

Maintaining clean, hygienic conditions that help prevent disease through services such as garbage collection and wastewater disposal.

(*Source*: *www.unesco.org/education/tlsf/theme_c/mod13/www.worldbank.org/depweb/english/modules/glossary.html*)

☆ A comprehensive term referring to the development and application of measures designed to protect public health. Sanitation procedures attempt to prevent the spread of disease through the proper disposal of sewage and other waste, correct handling of food and water supplies, and control of the carriers of disease (mosquitoes, rats, etc.).

(*Source*: *www.bfa.sdsu.edu/ehs/glossary.htm*)

Rural Youth

Rural youth is backbone every rural societies. Youth can be defined as:

Youth

Young person: a young person (especially a young man or boy).

(*Source*: www.wordnet.princeton.edu/perl/webwn)

- ☆ Young people ages 15 to 24 years.

(*Source*: www.rho.org/html/glossary.html)

Farm Women

Farm women are very important segment of rural society. Woman is defined as under:

- ☆ A woman is an adult female human being, as contrasted to a man, an adult male, and a girl, a female child. The term woman (irregular plural: women) is used to indicate biological sex distinctions, cultural gender role distinctions, or both.

(*Source*: www.en.wikipedia.org/wiki/Women)

Credibility

Credibility of farm information is very essential aspect of farm information communication. There are several definitions of credibility, which are as follows:

- ☆ The quality of being believable or trustworthy.

  (*Source*: www.wordnet.princeton.edu/perl/webwn)

- ☆ Credibility is the believability of a statement, action, or source, and the ability of the observer to believe the above.

(*Source*: www.en.wikipedia.org/wiki/Credibility)

- The trustworthiness (credentials, education, experience, etc.) of an author.

  (*Source*: *www.ucblibraries.colorado.edu/about/glossary.htm*)

- Defined by Webster's Collegiate Dictionary as the quality or power of inspiring belief. When applied to an information source, credibility centres on three factors: opportunity (is the source in a position to know?) ability (does the source have the skill and competence?) dependability (is the source responsible and trustworthy?).

  (*Source*: *www.inspection.gc.ca/english/corpaffr/publications/riscomm/riscomm_appe.shtml*)

Presentation

'Presentation' is important parameter of credibility of farm information which can be defined as:

- The act of making something publicly available; presenting news or other information by broadcasting or printing it; "he prepared his presentation carefully in advance".

  (*Source*: *www.wordnet.princeton.edu/perl/webwn*)

- Presentation is the process of presenting the content of a topic to an audience.

  (*Source*: *www.en.wikipedia.org/wiki/Presentation*)

Brevity

Brevity is a quality of a presentation which delineates to present the matter in brief and concise form. Brevity is defined as under:

- ✰ The use of brief expressions.
- ✰ The attribute of being brief or fleeting.

(*Source*: www.wordnet.princeton.edu/perl/webwn)

Detailed

Detailed presentation of any information helps the receivers in communication process. 'Detailed' can be defined as:

- ✰ Developed or executed with care and in minute detail; "a detailed plan"; "the elaborate register of the inhabitants prevented tax evasion"- John Buchan; "the carefully elaborated theme".

(*Source*: www.wordnet.princeton.edu/perl/webwn)

Understandability

'Understandability' is another important parameter of credibility of information.

Clarity

'Clarity' is the quality to present information in clear way. It is defined as under:

- ✰ Free from obscurity and easy to understand; the comprehensibility of clear expression.

(*Source*: www.wordnet.princeton.edu/perl/webwn)

Necessity

'Necessity' of information is a criterion which delineates how much the information is necessary to the receiver. The different definitions of necessity are given as under:

- ✰ The condition of being essential or indispensable.

(*Source*: www.wordnet.princeton.edu/perl/webwn)

- ☆ Something necessary or needed.

  (*Source*: *www.collections.ic.gc.ca/ghost/glossary.html*)

- ☆ Something that is necessary or needed or that cannot be done without.

  (*Source*: *www.ggca4thgrade.tripod.com/id10.html*)

Feasibility

'Feasibility' delineates the pragmatisms of information and it is important parameter of credibility. Different definitions of feasibility are:

- ☆ The extent to which a study or project may be done practically and successfully.

  (*Source*: *www.grants.nih.gov/grants/funding/phs398/instructions2/p3_definitions.htm*)

- ☆ A determination that a process, design, procedure, or plan can be successfully accomplished in the required time frame.

  (*Source*: *www.thequalityportal.com/glossary/f.htm*)

Universality

'Universality' of information is another important aspect which can be defined as:

- ☆ The quality, fact, or condition of being universal.

  (*Source*: *www.phy.ohiou.edu/academicpro/indepen/One/glos.htm*)

- ☆ The quality of being universal; existing everywhere

  (*Source*: *www.wordnet.princeton.edu/perl/webwn*)

Timeliness

'Timeliness' of information is such a character, which has an important role on credibility. It is defined as under:

- The time between the occurrence of an event and the availability of data about the event. Timeliness is related to the use of the data.

(*Source*: *www.intra.ccgh.com/AutomatedCAMH/gl.html*)

- Seasonableness: being at the right time

(*Source*: *www.wordnet.princeton.edu/perl/webwn*)

Current

Freshness of information is very important criterion to the receiver in a communication process, which is delineated as 'current'. It can be defined as:

- Occurring in or belonging to the present time; "current events"; "the current topic"; "current negotiations"; "current psycho-analytic theories"; "the ship's current position".

(*Source*: *www.wordnet.princeton.edu/perl/webwn*)

Repetition

'Repetition' of information helps the receiver to get the opportunity if he or she missed the information some how in communication process. Definitions of 'repetition' are:

- The act of doing or performing again

(*Source*: *www.wordnet.princeton.edu/perl/webwn*)

- Repetition is the occurrence of an event which has occurred before.

(*Source*: *www.en.wikipedia.org/wiki/Repetition*)

Biased

'Biased' is a very important parameter of credibility. It can be defined as:

- ☆ Influence in an unfair way; "you are biasing my choice by telling me yours".

  (*Source*: *www.wordnet.princeton.edu/perl/webwn*)

- ☆ In general, any factor that distorts the true nature of an event or observation.

  (*Source*: *www.gulflink.osd.mil/medsearch/glossary/glossary_b.shtml*)

- ☆ A preference or an inclination, especially one that inhibits impartial judgment.

  (*Source*: *www.student.biology.arizona.edu/honors98/group14/glossary.html*)

Accuracy

'Accuracy' of any information makes it more credible to the receiver and can be defined as:

- ☆ The quality of being near to the true value; "he was beginning to doubt the accuracy of his compass"; "the lawyer questioned the truth of my account"

  (*Source*: *www.wordnet.princeton.edu/perl/webwn*)

- ☆ The extent to which an evaluation is truthful or valid in what it says about a program, project, or material.

  (*Source*: *www.cdc.gov/tobacco/evaluation_manual/glossary.html*)

- ☆ A qualitative expression of closeness of the measurement to the true value.

  (*Source*: *www.xdcr.com/UT_terms.html*)

Factual

Any information based on facts increase its credibility. Different definitions of being factual are as under:

- ☆ Actual: existing in act or fact; "rocks and trees..the actual world"; "actual heroism"; "the actual things that produced the emotion you experienced".
- ☆ Of or relating to or characterized by facts; "factual considerations"
- ☆ Characterized by fact; "the factual aspects of the case".

(*Source*: *www.wordnet.princeton.edu/perl/webwn*)

Believability

'Believability' of information is another very important parameter of the credibility of information in a communication process which can be defined as:

- ☆ Ortony (2001) identifies "believability" as a way to gauge the accuracy of simulated human behaviour. We measure believability with a 1-5 Likert rating.

(*Source*: *www.openheartlogic.org/opine.cgi*)

- ☆ Credibility: the quality of being believable or trustworthy.

(*Source*: *www.wordnet.princeton.edu/perl/webwn*)

Theoretical Studies of Content Analysis

Probably the first quantitative newspaper analysis published in 1893, asked the rhetorical question, "Do newspaper now give the news?" by Speed (1893). Its author showed how, between 1881 and 1893, New York newspaper had dropped their coverage of religious, scientific, and literacy matters in favour of gossip, sports and scandals.

In 1903, Löbl published in Germany, an elaborate classification scheme for analyzing the "inner structure of content" according to the social functions that newspapers perform.

Street (1909), by simply measuring thee column inches that newspaper devoted to particular subject matter attempted to reveal "the truth about newspapers".

In a far more simplistic study published in 1910, Mathews attempted to reveal the overwhelming space that one New York daily newspaper devoted to "demoralizing", "unwholesome" and "trivial" matters as opposed to "worthwhile" news items.

At the first meeting of the German Sociological Society in 1910, Weber (1911) proposed a large-scale content analysis of the press, but for a variety of reasons the research never got off the ground. During the same period, Andrei Markov (1913), who was working on a theory of chains of symbols, published a statistical analysis of a sample of Pushkin's novel in verse, 'Eugene Onegin'. These inquiries were discovered only recently or influenced the content analysis literature only indirectly.

Lasswell (1941), in his symbol analysis led to his "World Attention Survey" where he compared trends in the frequencies with which prestige newspapers in several countries used national symbols.

Janis (1943) excluded pragmatic content analysis because inferences as to the causes or effects of content can rarely be validated solely by analysis of messages themselves.

On the other hand, proponents of a broader definition usually assume that content data will be used, directly or indirectly, in conjunction with independent behavioural indices.

The first concise presentation of the conceptual and methodological developments under the new umbrella, term content analysis appeared in a mimeographed text titled 'The Analysis of communication content authored by Berelson and Lazarsfeld (1948), which was later published as Berelson's Content Analysis in Communication Research (1952).

According to Berelson (1952), content analysis is a research technique for the objective, systematic, and quantitative description of the manifest content of communication. Berelson (1952) defined it as less restrictive and include studies in which findings are reported in such terms as "more", "less", or "increasing". Berelson (1952) also suggested the method of contention analysis is used for analyzing the characteristics, causes and effects of content.

Despite the widely recognized advantages of quantitative methods, the tendency to equate content analysis with tabulation of frequencies has been criticized on a number of grounds. The most general of these is the charge that such a restriction leads to bias in the selection of problems to be investigated, undue emphasis being placed on precision at the cost of problem significance according to Barcus (1959) and Smythe (1952).

Probably the first computer aided content analysis was reported by Sebeok and Zeps (1958) who made use of simple information retrieval routines to analyze some 4000 Cheremis folktales.

Pool (1959) reported that it should not be assumed that qualitative methods are insightful and quantitative ones merely mechanical methods for checking hypotheses. The relationship is a circular one–each provides new insights on which the other can feed.

Osgood (1959) reported that the interest of the content analyst lies in making inferences about the source of a message, he must rely upon encoding dependencies, that is, the dependencies of message events upon psychological processes in speakers and writers. When his interest lies in making inferences about the effects of a message upon its receivers, on the other hand, he relies upon decoding dependencies, that is, the dependencies of events in listeners and readers upon the content and structure of messages.

George (1959) reported that qualitative analysis of a limited numbers of crucial communications may often yield better clues to the particular intentions of a particular speaker at one moment in time than more standardized techniques. Qualitative content analysis, which has sometimes been defined as the drawing of inferences on the basis of appearance or non-appearance of attributes in messages, has been defended largely, though not solely, for its superior performance in problems of applied social science.

Hays (1960) explored the possibility of designing a computer system for analyzing political document.

In 1974, participants in the 'Workshop on Content Analysis in the Social Sciences', held in Pisa, Italy saw the development of suitable algorithms for computer content analysis as the only obstacle to better content analysis (Stone, 1975). Since that time computational approaches have moved in numerous direction.

Namenwrith and Weber (1987) opined that–one reason content analysis did not blossom, the survey research did was that the lack of standardized procedure and measuring instrument worked against the accumulation of comparable results.

Weber (1990) remarked, "There is no simple right way to do content analysis. Instead, investigators must judge what methods are most appropriate for their substantive problems. Moreover, some technical problems in content analysis have yet to resolve or are the subject of ongoing research and debate.

According to Riffe *et al.* (1996) a monthly stratified a sample of 12th issues is the most efficient way to assess news magazine content. A random sample of 16th issues permits somewhat more accurate inferences about content, but this sampling method is less efficient.

Riffe *et al.* (1996) contended that the method of coding only a sample of all coded units by multiple coders in many content analyses possess potential problems that have seldom been discussed in the extent literature they provided a formula designed to estimate sample sizes required for valid reliability assessments.

Popping (1997) found that researchers who conduct network analysis of texts must do more to take into account the three sources of ambiguity–idiomatic ambiguity (*i.e.* the presence of idiomatic words and phrases), elocutionary ambiguity (in which the elocutionary intention of a sentence is unclear) and relevance ambiguity (in which some relationships between themes may be irrelevant).

Hagedorn *et. al* (1998) developed a frequency index (based on CH 12 and Tukey's 'W' test) to compare content analysis results and develop ranking as a measure of the relative importance of each issue of popular press and technical or regulatory on public perception of transgenic in bio-control.

Pal *et. al* (1999), also acknowledged that personal localite sources are more important to the farmers than mass media. Although there is much scope for dissemination of techniques through mass media sources.

Potter and Levine-Donnerstein (1999) found that "Validity and reliability should be conceptualized differently across the various forms of content and the various uses of theory "found in contemporary content analysis". They identified 3 types of content-manifest, Latent, Pattern and Projective that differ in the extent to which coders are asked to apply personal interpretative scheme in coding decision.

Study Related to Farm Information Disseminated through Mass Media

Rogers and Beal (1958) reported that mass media like farm magazines were most often mentioned as source of information by the Ohio-farmers. He further stated that mass media were generally most important in creating awareness of new ideas but personal influence of neighbours and friends was regarded as most effective in convincing farmers to actually try out the new ideas.

Berlo (1960) stated that magazine readers have an image of the magazine they read. They expect different messages, and, message treatments from different magazines.

Peterson *et al.* (1964) speculated that the modern magazines of low price, popular appeal and large circulation emerged in the last decades of the nineteenth century.

Ernst (1972) conducted an interesting study with three newspapers published at national, state and country levels. The coverage items in these newspapers indicated that at all the levels the sports news received higher preference over news pertaining to community development.

Newspaper is one of the effective mass media in any country. In developing society "change accepting personalities" arise and they become radio listener, newspaper reader etc. Chiders and Vajrathan (1973) indicated that the mass media, especially the printed media, are extremely important in stimulating a sense of environment.

Dey (1976) wrote, "in comparison with the other media Newspaper is the oldest mass medium". Radio and Television have created significant influence on the society. In some fields of communication the Radio and Television have crossed the scope of Newspaper.

Naccarato and Kimberly (1998) found that the content of business to business magazine advertising can be used to predict readers' recall and evaluations of advertisement. Several "form" variables (*e.g.*, use of colour) and "content" variables (*e.g.*, use of fear appeals) were found to be statistically significant predictors of readers attention, recall and evaluation.

Pal *et. al* (1999) revealed that personal localite sources are more important to the farmers than mass media. All though there is much scope for dissemination of techniques through mass media sources, the preferences of farmers must be taken into account in their design.

Bharati and Sohi (2001) reported that exposure of the farmers to mass media was significantly associated to their use of level of rice technology.

Study Related to Content Analysis of Farm Information

Mathur *et al.* (1978) conducted a nationwide survey in 1977 on a sample of 103 farm journal editors and 103 publishers through separate mail questionnaires. The major findings were (i) Readership was in rural areas of India, (ii) Circulations were more in farm families, extension workers and scientists, (iii) the Government and teaching-cum-research organisation included research based articles but individual publishers included success stories and news articles more than others, and (iv) the sources of information were teaching and research organisations for government and teaching organisational publication but for private organisations it is still a problem to have good articles.

Nanjappa and Ganapathy (1987) compared three selected Kannada dailies in terms of space provided to different categories of agricultural information, agricultural news category occupied the maximum space (31.17 per cent of total agricultural information). All the information published on agriculture was categorized under a different heads such as (i) Agricultural news, (ii) Market price, (iii) Advertisement, (iv) Letter to the editor, (v) Feature articles,

(vi) Editorial, (vii) Suggestion to farmers, (viii) Weather, (ix) Questions and answers and (x) Success stories.

Sawant and Shendre (1987) reported that the newspapers provided space for agricultural information (2.32 per cent) and agricultural advertisement (1.87 per cent) was less than non-agricultural information (38.68 per cent) and non-agricultural advertisement (57.16 per cent) in their study on Marathi newspaper 'Sakal' from Maharastra.

In a study on regional German farming newspapers McHenry (1996) found that there were many interwoven arguments, which included traditional agricultural discourses as well as the newer environmental ones. It was clear that different aspects of agricultural and environmental issues were highlighted where they appeared to be beneficial to farmers' interest. Further more, it was found that in the promotion of the use of renewable resources, which was portrayed as being in the interest of farmers.

In the study on the programmes of farm and home broadcast in Radio, Sasidhar *et. al* (1999) found that nearly half of the programmes were broadcast on agriculture followed by horticulture, animal husbandry, extension, Farmwomen, sericulture and developmental programmes. Time allotment to the different subject matter areas as that of topics covered. Interview format was used for broadcasting majority of topics.

Velusamy and Seetharaman (2000) conducted content analytic study of four dailies to know the extent of coverage of subject matter areas, source of articles modes of presentation and illustration components. The results depicted that higher percentage of articles were from agriculture as scientific articles from University Scientist with less percentage illustration components.

Caldwell et. al (2000) studied media coverage of integrated pest management in major urban newspapers in U.S.A. from 1982 to 1995. They reported that overall numbers of articles published annually in each newspaper were relatively low. Much of the content dealt with urban settings and did not serve to improve understanding of the role of IPM in agriculture.

Nagabhushanam and Nataraju (2000) conducted a content analysis of agricultural information published in 'Prajavani', a Kannada daily in Karnataka, India. Results showed that the maximum number articles published were on horticulture (34) and field crops (23), with photographs extending up to 33.45 per cent and 21.81 per cent respectively. The majority (57.27 per cent) of the articles were contributed by progressive farmers, staff correspondents and persons interested in agricultural education.

Fawole and Olowu (2000) conducted a content analysis in Nigeria from 1970 to 1990 and reported that the public newspaper reported more of women's agricultural activities (63.2 per cent) than that of private newspapers (36.8 per cent). Large newspapers covered more (57.4 per cent) of women's agricultural activities than small (42.6 per cent) newspapers. It was also found that size and location significantly influence the coverage of women's agricultural activities by newspapers.

Singh (2001) conducted a content analysis of farm broadcast programme on All India Radio, Imphal on a period of 30 days (1 September 98 to 30 September 98). He reported that 42.50 per cent of total time slot for farm programme occupied by folk plays, folk music, folk songs and short radio plays. Other programmers, which followed were agricultural (11.65 per cent), general (10.55 per cent) and horticulture crop protection was broadcasted in greatest numbers (33.33 per cent), where as least time slot was devoted to crop nutrition (7.87 per cent).

Yiping *et. al* (2002) reported that number of readers for each issue of farm journals were highest in the 1950s and 1960s in Japan, but have decreased as the numbers of journal increased.

Adigun (2002) analyzed 348 publications written by research scientist at the West Africa Development Association (West Africa) from 1991 to 2000 and found that rice agronomy (22.7 per cent), soil science (19 per cent) and rice breeding or genetics (14.9 per cent) were the most covered disciplines in rice science.

A study of content analysis of 1800 editions of randomly selected daily newspapers published in Nigeria from 1st January 1985 to 31st December 1994 was conducted by Oladeji and Olowu (2003). They investigated the information cues of agricultural advertisements. The results revealed that majority of the agricultural advertisements have product prices and location, product utility and product description while product capacity and product process were rarely used. Also specific information cues were prevalent in the advertisements of particular agricultural products and services and there was no significant variation in the information cues of agricultural advertisements overtime. However, there were indications that farmers may likely obtain necessary information that may benefit them from the advertisements of agricultural products and services.

Zhu *et al.* (2003) searched more than 20000 papers on tea science from all accessible resources using the most advanced technologies of network, information and database. The results showed that 21206 papers were published in the last decade or so, distributed in 2073 publishing media and involving 9132 authors who belong to 6697 institutions.

Roy *et al.* (2003), conducted a content analysis on five leading Bengali dailies, *Ananda Bazar Patrika, Bartaman, Sambad Pratidin, Aajkal,* and *Ganashakti* for the period of two months, starting from 16th Oct' 1999 to 15th Dec' 1999. It was found from the study that all the five newspapers were not giving adequate coverage to farm news. The coverage of farm news was some what satisfactory in *Ganashakti, Bartaman* and *Sambad Pratidin* while it was altogether unsatisfactory in *Ananda Bazar Patrika* and *Aajkal*. Marketing aspects, natural hazards, agricultural polity and social life received more importance in those newspapers.

Nieckerson and Hellerstein (2003), in an extensive content analysis of the enabling legislation of various farmland protection programme suggested that wide variation exist in the protection amenities. It also revealed unique preferences regarding the spatial patterns of preserved agricultural lands.

Marks *et al.* (2003) conducted a study on United Kingdom and United States newspapers over the period 1990 to 2001 to examine media coverage of genetically modified (GM) crops in a risk communication framework. Findings showed the UK press has been more negative than the U.S. press in its coverage of GM crops. In addition, environmental and food safety events had a significant impact on the level and cycle of GM crop coverage.

Reisner (2003) in his study examined how six national newspapers in the USA balanced supporting agriculture (a morally good occupation) with supporting environmentalism (nature as a moral value), in an area in which agricultural and environmental interest conflict (farm use of pesticides). Data are based on each newspapers indexed for the year 1988 to 1990. Farmers were portrayed positively as quiet social movement participants, and newspapers suggested that government and universities were blocking infrastructural change that should be supported.

Maris and Roe (2004) conducted a study to investigate public opinion on automatic milking. A descriptive content analysis of newspapers from six countries (Belgium, Denmark, Germany, The Netherlands, Sweden and U.K) was conducted from November 1, 2001 to October 31,2002 to assess the ways in which newspapers covered agricultural and food issues. A total of 1385 news were recorded in the 14 European newspapers. Out of the 1385 news item concerning agriculture and food chain, only 3 dealt with automatic milking. 41.3 per cent of all items were evaluated as negative, compared to 21.5 per cent positive. It is also implied that the development was not an issue and was perhaps best left alone.

Hudeckova and Lostak (2004) outlined the experiences with SAPARD programme as they are recorded in Czech print media (newspapers and specialized journals) published since the beginning of 2002 up the October 2003. It is shown that it is the rural stakeholders who are well prepared to act in the institutionalized frames of the EU structural policy. The issues that have to be addressed include: better coordination of activities; a good system of information; dissemination of gathered experiences and simplification of administration.

Study Related to Media Credibility

Readability and comprehension tests provided suitable evidences to writers how to write with comprehension and how to improve the understandability of the printed material. Researches on readability tests started after 1930 and through careful research various readability and comprehension tests had been developed by Flesch (1948), Gunning (1952) and Taylor (1953).

Flesch (1948) used average number of syllables per hundred words and average sentence length in words in his formula.

Ensminger (1957) recommended that publication must be written in short sentences and in simple words. He also suggested the use of attractive sources and easily understandable photographs.

Berlo (1960) viewed the linguistic facility of communication source and its compatibility with the receiver's understanding acted as an important factor in communication process and its high fidelity.

Fox (1963) emphasized contents prerequisites in the style of writing of agricultural and technical articles so as to make it comprehensible to the intended reader. He suggested that clarity and brevity in writing are important for better comprehension and called for simple and short sentences with crisp and concrete words.

Kelsey and Hearne (1965) recommended that the periodical should have a cover page of heavy material, illustration be placed near the test reference and in natural. They also suggested using the convenient size for the pamphlet and other periodicals.

Kammath (1969) prescribed that the message in an agricultural article should be related to some problem or ambition of the reader and it should be properly timed. The message should be related to immediate future readers, so as to attract readers' attention.

Kammath (1969) pointed out that even unfamiliar technical words also could be used in farm story writings provided they are well interpreted and explained with familiar words. Relevance of the message can be considered as an important prerequisite to get news value.

Patel and Patel (1972) studied effectiveness of "Khadut Patrika" in Gujarat. They emphasized upon having attractive cover pages, attractive printing of cover design with combination of two or more colours along with own some action pictures. For second, third and fourth cover pages they suggested printing of slogans, key points and message etc. They also recommended the use of attractive pictures along with the related text. They asked the writer to avoid the use of even common English words in writing articles for farmers.

Flesch (1974) concluded that human interest makes reading easier. Scientific tests have shown that people are better at reading about other people, than about anything else.

Somesundaran and Jagannath (1974) reported, therefore, it is advisable to test the readability of any farm literature produced before its publication.

Kayal (1975) in his study of a Bengali farm magazine "Krishi Pragati" in West Bengal found that the level of education of the respondents had significant relationship with readability as well as comprehension of articles. With the increase in educational levels, the Illustration comprehension scores of the respondents also increased. The readability of content was highest for readers belonging to medium age group (52 per cent) and lowest in young age group (12 per cent). With regard to coverage more space was devoted to the areas of crop production, vegetables and fruits cultivation and agricultural news respectively.

Dey (1976) opined that the importance of newspapers in respect of other mass media may be classified on the following points:

1. *Credibility*: It is an acceptable fact that Newspapers is the oldest medium of mass communication. During last three centuries many talented personalities have come in contact of mass communication and development of this media and made credibility of it.
2. *Fighting Tradition*: In ancient days of mass communication the editors and Publishers of Newspapers had to fight against rigid decision of the management. May be some of them had to lose their lives even. Many had to bear oppression. For

many reasons the Newspaper earned respect and reputation from the public.

3. *Information bias*: Other mass media are mainly based on entertainment but the Newspaper on news and truthfulness of fact and reality. So in fact, news for and against, analysis of reason of fact, observation, etc. as found in Newspaper have created respect and eagerness. A farmer agrees to invest against a crop seeing its profitable result written by reputed scientists.
4. *Portability*: Newspaper can be carried easily from one place to other. It can be read in leisure hours even can be preserved easily. A farmer can read repeatedly the dozes of plant protection chemicals, growth hormones or the seed rate and important information even at the time of application.
5. *Wide message choice*: As the Newspaper is a package so the reader has got the right to choose news item of his own. These advantages are not found in Radio and Television.
6. *Private contact*: In Democratic countries Newspapers are owned by the Private Sector. So these newspapers can write Government free opinion. In modem age freedom of press is an acceptable fact. As the investment for publishing a small or medium sized newspaper is small. So a man can easily publish the same.

Kaur and Mathur (1981) reported that the content of a farm magazine is by far the most important determinant of its success and only an attractive cover page and format will not help if the content is not timely, need based and locally relevant.

Das (1987) stated considering farm journal as a mass media, it is an instrument for effective communication to the farmers. The responsibilities of the reporters and the editorial board of the journal or the newspaper, therefore, is to report what is latest in technology and how far the information could be effectively utilized by the readers for improving the status of farming.

Patel *et al* (1995) found that source credibility of radio and newspapers in progressive villages are 1.085 and 0.192 respectively, whereas in less progressive villages they were 0.399 and 0.160.

Angadi and Swamy (1995) conducted a study on credibility of Agricultural Assistants as perceived by farmers was measured using a scale which consisted of five components (safety, qualification, communication competence, personality and homophily). They were rated fairly high on main components such as safety and qualification, moderately on homophily and communication competence and poorly on personality.

Schwarz and Doluschitz (1997) reported that timeliness was particularly valued in some areas such as crop protection. Electronic media were the preferred method of access for information that was either time critical or of local importance.

Schweiger (1998) found newspapers in Germany were rated ahead of the web and television on nine of eleven credibility item.

Kiousis (1999) found news credibility perceptions to be influenced by media use and interpersonal discussion of news. He found general skepticism about news, but people rated newspapers as more credible than online news or television.

Flanagin and Metzger (2000) noted that newspapers, books and television undergo a process of information verification before they reach the public.

Kalita and Singh (2003) reported that village level workers were most credible sources of information. Farmers have negligible preference to both print and non-print mass media.

Study Related to Farm Information Disseminated to the Farmers

Riley and Riley (1959) interpreting the findings of Lazersfeld stated that the use and perhaps the effectiveness of different media varied with the educational level of audience; the higher the education, the greater the reliance of print, the lower the education the greater the reliance on rural and picture media. They further interpreted the findings of Lazarsfeld and stated that the audience in fact overlapped to a marked extent: that the movie goers were also radio-listeners, magazine readers were also T.V. fans and nearly all of those at least looked at one or more newspapers fairly regularly.

In 1950 and 1955, in Washington County, Kentucky, 285 farmers were interviewed twice. Farmers reported using farm magazines (70 per cent), newspapers (58 per cent) and farm bulletin (42 per cent) on both the surveys (Conghnour, 1960).

Rai (1965) observed that higher the education of the farmers, greater was the interest in reading various kinds of literature. Agricultural magazines were least preferred by the farmers. Size of holdings and education were positively related to information seeking habit.

Ray (1967) found that high and low urban groups had significantly different participation in newspaper reading. Not only more respondents from high urban group participated in newspaper reading, but they participated more frequently.

In a report 'Agro- information flow' at the 'village level' (1968) by Indian Institute of Mass Communication, it has been found that among the mass media, newspaper was desired source of contact next best to radio and it did serve as first source of information for nearly 3 per cent of the farmers interviewed and 1.33 per cent as subsequent source of information (interest stage). It is stated that man's language affects his perception and his thinking.

In days to come we should focus more study on the use of written words. The myth of non-applicability of written words due to illiteracy of farmers has been punctured by a study of Singh and Nirwal (1974). The author based on audience analysis for using written words in a Hariana village reported "when we use the literacy score of head of the farm families alone, only 42.47 per cent of the total farm families make potential audience for written words, when the literacy score of the family as a whole is considered, this figure goes upto 78.95 per cent, considering 'FLS-IN' only, it comes to 78.07 per cent. When we also consider the farmers' linkages in terms of their expressed effective or instrumental preferential, it rises to 91.22 per cent when we also encompass the family literacy position of the expressed preferential; it rises to 94.73 per cent. This shows that even though majority of farmers may be illiterate, he has somebody in his immediate environment to read out the written words to him.

Khandekar (1975) studied 'Unnat Krishi' a monthly journal published by Farm Information Unit of Directorate of Extension and found that 60 per cent of the readers had farming as their main occupation-, whereas 38 per cent had service or business as their profession besides farming. As regards the effectiveness of 'Unnat Krishi', fifty per cent of the readers perceived the magazine to be effective, 32 per cent thought it to be most effective, whereas only 18 per cent perceived it to be less effective. Hence, it can be concluded that 'Unnat Krishi' was found to be effective. It was also noted that perceived effectiveness of the magazine seemed to increase up to the age of 25 years and declined after 45 years. The education, mass media exposure, reading ability, comprehension skill and functional literacy were positively related to the perceived effectiveness of the magazine.

Khandekar (1975) also found out that the areas on which the readers wanted information in a farm magazine were crop production, Animal Husbandry and Dairy and Vegetables and Fruits cultivation.

Khandekar and Mathur (1975) conducted a study on effectiveness of 'Unnat Krishi' for different categories of readers. He found out intimate correlation between specific age groups and educational groups and effectiveness of 'Unnat Krishi'. He also found out that effectiveness of farm magazine like 'Unnat Krishi' varies with functional literacy levels and extent of mass media exposure of the readers. He further emphasized upon having an in built mechanism for getting feedback from the readers.

Sawant (1982) reported an analytical study of the advertisements. Farmer readers perceived heading dominant advertisement be more effective, whereas illustration dominant advertisement was perceived as more effective by farmer non-readers. Advertisement with all elements emphasized and printed in colour was perceived more effective. The communicability of heading dominant advertisement and heading and illustration dominant advertisement were the same.

Kota (1983) in a study of readership and effectiveness of farm magazine 'Anna Data' observed that nearly 10 per cent of the respondents read only head lines and titles, 52 per cent read the magazine issues partially and 38 per cent of them reported reading the 'Anna Data' issues fully.

Kumar (1983) carried out a readership survey of 'Kisan Bharati' a farm magazine brought out by Pantnagar. The magazine had a 2,630 subscriber in the month of August, 1982. Mostly, subscribers were of middle-aged belonged to higher castes and had larger land holding, often being published irregularly and material comprehensive irrespective of age.

Ansaloni (1986) showed in a survey in Emilia-Romagna on livestock production, the information gained from all the media: 86.7 per cent watched TV daily, 59.1 per cent read a specialized publication and 41.8 per cent listened to the radio daily.

Mishra and Tripathy (1991) found in their study of exposure of farmwomen to different sources of information that most of the respondents listened to the radio and exposure to printed media was minimal.

Silva and Garforth (1997) stated that the interview format is the most effective in terms of immediate knowledge gain, followed by drama, discussion and talk form among the different radio programme format for the dissemination of information in Srilanka.

Ani and Kwaghe (1997) in their study found that radio is the most frequent source of form information. Significant associations were reported between the use of radio and factors such as age, length of farming experience and net annual farm income. It is suggested that the use of radio as a medium of reaching farmers should be intensified by the extension agencies.

Rodriguez *et al.* (1997) reported that when full time and part time Iowa farm operators and their families are asked where they obtained safety information, 95 per cent of the respondents said newspapers and magazines. 82 per cent radio, 77 per cent television, 59 per cent relied on publications from ISU Extension and 33 per cent relied on ISU Extension staff.

Omotayo *et. al* (1997) in their study on Nigerian farmer found that 28 per cent of the respondents across the country owned a radio set, while about 48 per cent had access to one. Across the country, respondents indicated the evening time as the most appropriate for radio farm broadcast.

Sarkar (1997) stated that radio, television, newspapers and leaflets were the most frequently used sources, while campaigns, demonstrations, periodicals and farm magazines were less frequently used by most of the respondent in Ranaghat- II block of Nadia district, West Bengal, India.

Ladebo *et al.* (1997) indicated that the use of the radio by the farmers is not dependent on their socio-economic characteristics. However, the radio should be complemented with other media for information dissemination in realizing optimum benefits from the efforts of the extension service.

Reddy and Reddy (1997) reported that farm size, mass media exposure other than radio, socio-political participation, extension participation activity, economic orientation and extension contact had greatest influence on listening behaviour of farmers towards farm broadcast.

Jondhale and Nandapurkar (1998), reported that subscribers differed on the readability, credibility, understandability, brevity, directness, illustrativeness, imaginativeness, enjoyment in reading in a study of utility of articles published in 'Shetibati' and 'Shetkari' journals on the basis of 160 samples and fourteen variables.

Das and Sharma (1998) revealed that farmers accorded higher credibility to personal sources of farm information than to impersonal sources.

Kashem (1999) found that radio ranked third in terms of contact with top eight information sources by the farmers and prioritization of preferences in Bangladesh while in Japan newspapers ranked second.

Rahman *et al.* (1999) revealed that among the different sources of information, radio ranked fourth in terms of most credible sources in disseminating HYV Boro rice technology as perceived by the block supervisors.

Vannirajan *et. al* (2000) studied readers of the newspapers "Dinamalar" and "Dinamani" in 7 village in Madurai and results showed that farmers preferred subject areas related to agriculture, horticulture and agricultural economics.

Egbule and Njoku (2001) revealed that among the other things that the mass media have performed poorly in disseminating requisite agricultural information to farmers, although there is a positive correlation between mass media usage and farm yield. The greatest percentages of rural farmers receive information on modern agricultural practices from radio, although they prefer television.

Kumar and Se (2001) observed that most preferred mode of communication in farm radio programme among the respondents was interview method. It was preferred by 76.6 per cent of the respondent. The next preferred mode was Drama (68.8 per cent). The third and fourth preferred modes are folk songs and straight talk.

Oyegbami and Fabusoro (2003) examined the use of radio and television as source of agricultural information among poultry farmers in Egbeda local government area, Nigeria. They found that the respondents' level of use of these media was on the average. The average age and years of formal education of the respondents were 38.7 years and 10.8 years respectively. Also 37 per cent had a very high level of cosmopoliteness. A significant relationship was also established between respondent's level of production and their level of use of radio and television at 5 per cent confidence interval.

Statistical Approaches on Credibility Study

Factor analyzing word counts to infer themes have a long history, which is reviewed by Iker (1974). Stefflre (1965) is an early pragmatic statement. Moreover, each approach entails a different measurement model. Specifically, first order exploratory factor analysis is the statistical model that corresponds to inferring themes from word co-variation. The measurement model for single classification assumed dictionaries corresponds to a restricted second confirmatory factor analysis (Namenwrith and Weber, 1987).

Kim and Mueller (1978) viewed that factor analysis may be used as on expedient way of ascertaining the minimum number of hypothetical factors that can account for the observed co-variation and as a means of exploring the data for possible data reduction. This form of use is exploratory.

Kim and Muller (1978) suggested four basic steps of factor analysis:

1. The data collection and preparation of the relevant co-variance or correlation matrix,
2. Extraction of the initial factors,
3. Rotation to a terminal solution and interpretation, and
4. Construction of factor scales and their use in further analysis.

They also suggested that in exploratory factor analysis, one may rely on the use of correlation matrix.

Chapter 3

Methods and Cooperation

Selection of Mass Media

To study the farm information communicated through mass media, three mass media were selected initially. They were four Bengali newspapers, three Farm and Home broadcast by *All India Radio, Kolkata* and Farm and Home telecast by National Channel (*Doordarshan*). But after analyzing some basic facts, Television has been omitted from the study. Before conducting the final study, a pilot study has been conducted on ninety respondents (thirty farmers randomly selected from each three sub-divisions) to understand the basic situational background of the respondents. The Table 1 will justify the reasons for omitting Television, which was found during pilot study.

Multiple choices were found in the pilot study from the respondent in terms of their assets possession, media access and preferences of media as the sources of gathering farm information. It was found in the table that only 5.56 per cent of the farmers have Television of their own and only 8.89 per cent have access to Television where as out of ninety farmers, most of them have access to Radio and

Table 1: Distribution of Preferences of the Respondents as the Sources for Obtaining Farm Information

Media	Subscribe or Possess (Out of 90 Respondents)		Have Access (Out of 90 Respondents)		Preferences of Media	
	Frequency	Percentage	Frequency	Percentage	Frequency	Percentage
Newspapers	44	44.89	72	80.00	66	43.42
Radio	57	63.33	85	94.44	79	51.97
Television	5	5.56	8	8.89	7	4.61
No As Such Assets	4	4.44	2	2.22	0	0
Total	–	–	–	–	152	100

Newspapers. Table 2 depicts that Radio ranked the highest (51.97 per cent) and Newspapers (43.42 per cent) was next to it in terms of preference of media for getting farm information. Only a very negligible (4.96 per cent) cases Television was found as a source of farm information. Farmers opined that their socio-economic condition is not good enough to have a television set in their home. They also responded that some private channels disseminate good quality of farm information, but most of the farmers have no financial strength to access these private pay channels. National Channel (*Doordarshan*) is very recently telecasting Farm and Home Programme in regular basis as a scheme on Mass Media Support to Agriculture Extension from 2003-2004. During the data collection stages of the study, National Channel (Free Channel) had also disseminated farm information, but they are very irregular. Due to these reasons farmers were not interested for gathering farm information from the Television. After analyzing these basic facts, Television has been omitted from the study, as it has no practical utility and relevance for the farming community of the study area.

Content Analysis of Farm Information Communicated through Newspapers

A. Universe of the Study

The four most widely circulated Bengali newspapers have been selected as the universe of the study. These four newspapers have been widely circulated in every part of West Bengal and also to the Bengali-speaking people in India and abroad as well. The four Bengali newspapers are:

- ✰ *AAJKAL*
- ✰ *ANANDA BAZAR PATRIKA*

- *BARTAMAN*
- *SAMBAD PRATIDIN*

A complete enumeration of one year period has been done regarding selection of farm information published in the above-mentioned four Bengali newspapers for the analysis of the content of farm information. Before deciding upon the actual study period, different cropping seasons of the study area (West Bengal) have been considered. West Bengal in general and Birbhum district in particular has three distinct cropping seasons, namely, *Kharif* (Monsoon: June–October), *Rabi* (October–February) and *Boro* (February–April).

B. Categorization of Farm Information

After careful study of all the farm information published in all the four selected newspapers during the period are categorized into three major categories:

- NEWS
- VIEWS
- ADVERTISEMENTS

Different categories are again separately divided into different sub-categories according to type of information published during the period of study.

News and Views categories of farm information are divided into same categories. They are:

- Agricultural Pollution
- Agricultural Policy
- Agricultural Crisis and Management

- ☆ Agricultural Marketing
- ☆ Natural Calamities
- ☆ Rural Development
- ☆ Social Conflict
- ☆ Social Forestry
- ☆ Success Story
- ☆ Technology Transfer
- ☆ Weather Report

Similarly different sub-categories considered for Advertisement categories of farm information are:

- ☆ Agricultural Machinery
- ☆ Cooperative
- ☆ Fertilizer
- ☆ Plant Protection Materials
- ☆ Rural Development
- ☆ Seed
- ☆ Fishery
- ☆ Social Forestry

C. Reliability Test of Categories and Sub-categories of Farm Information Communicated through Newspapers

Reliability test for the process of categorization and sub-categorization is very much essential in content analysis, which helps in understanding the extent of consistency and reliability of the subject under study. For the present study, Krippendorff's Alpha (1997) has been calculated for establishing the reliability of the categories and sub-categories done. For the purpose, researcher and two other observers were taken. Observed disagreement and expected disagreement were measured. The decision rule is that when alpha is 0.8 and above, the subject under study is considered reliable. The value of alpha for different categories and sub-categories of farm information communicated through newspapers was found 0.832. Hence, the process of categorization and sub-categorization of farm information published in different newspapers was found reliable.

D. Space Covered

Amount of space provided for publishing any information in newspapers is an indicator, from which importance given by publishing house on the particular item can be ascertained. For the present study space covered by the farm information has been calculated in Sq. cm. and are summated for a year. Total Space published by the individual newspaper has been calculated by measuring the area of every single page for one year period.

Space covered by individual farm information is categorized into ten different strata for measuring frequency distribution. These are as follows:

Space Covered		*Assigned Code*
Up to 50 sq. cm.	-	1
Above 50 sq.cm. to 100 sq.cm.	-	2
Above 100 sq.cm. to 150 sq.cm.	-	3
Above 150 sq.cm. to 200 sq.cm.	-	4

Space Covered		Assigned Code
Above 200 sq.cm. to 250 sq.cm.	-	5
Above 250 sq.cm. to 300 sq.cm.	-	6
Above 300 sq.cm. to 350 sq.cm.	-	7
Above 350 sq.cm. to 400 sq.cm.	-	8
Above 400 sq.cm. to 450 sq.cm.	-	9
Above 450 sq.cm.	-	10

E. Title Font Size

Title Font Size of the farm information is another very important indicator, which indicates the relative importance given to the different farm information by the newspapers. Larger the font size, higher the importance. For the present study title font size of all farm information published were measured in order to ascertain the relative importance given by the publishing house towards farm information. Title font size is measured with help of existing Bengali Software Packages (iLEAP, Bangla Word) used by different newspapers. Title font sizes are categorized into different strata for measuring frequency distribution.

Font Size		Assigned Code
Up to 10	-	1
Above 10 to 20	-	2
Above 20 to 30	-	3
Above 30 to 40	-	4
Above 40 to 50	-	5
Above 50 to 60	-	6
Above 60 to 70	-	7
Above 70 to 80	-	8
Above 80 to 90	-	9
Above 90	-	10

F. Readability

Readability is another very important dimension of any print material, which signifies the acceptability of any information in printed form. Readability has been measured by Fog Index developed by Gunning (1952). Higher Fog Index, less the readability of any print material. The technique involved the following steps:

- ☆ Counting the number of words in a sample.
- ☆ Counting the number of sentences in a sample.
- ☆ Calculating the average number of words per sample.
- ☆ Calculating the average number of hard words (hard words–all the words with three or more syllables, abbreviations and symbols. Capitalized words unless symbolized or abbreviated are not counted).
- ☆ Calculating the percentage of hard words in a sample.
- ☆ Summing of word average and hard word percentage.
- ☆ Calculating Fog Index with the help of the formula:

  Fog Index = 0.4 (X+Y)

  where,

  X = Average number of words per sample,

  Y = Hard word percentage in the sample.

Fog Index has been calculated for each farm information published in different newspapers and then mean values have been calculated for every newspaper.

Content Analysis of Farm Information Communicated through Radio

A. Universe of the study

Farm Programme from All India Radio, Kolkata at 456.6 m Band and 657 KHz are studied and recorded. Three Farm Programmes have been identified as the universe of study which is broadcasted almost everyday in Bengali language. These are as follows:

- ✰ *CHASI BHAIDER BOLCHHI* (Morning)
- ✰ *UNO JAMIR DUNO PHASAL* (Afternoon)
- ✰ *KRISHI KATHAR AASAR* (Evening)

Like newspapers, a complete enumeration of one year period has been done regarding selection of farm information broadcasted by *All India Radio, Kolkata* for the analysis of the content of farm information.

B. Categorization of Farm Information

After careful study, the farm information disseminated by radio are categorized into four major categories on the basis of mode of delivery. These are:

- ✰ DELIBERATION
- ✰ INTERVIEW
- ✰ RADIO PLAY
- ✰ PANEL DISCUSSION

All the major categories are again separately divided into different sub-categories according to type of information broadcasted during the period of study. Accordingly, 'Deliberation' has been sub-categorised as follows:

Similarly, different sub-categories considered for the farm information broadcasted through Deliberation are:

- ☆ Agricultural Policy
- ☆ Crisis
- ☆ Farm Women
- ☆ Health and Sanitation
- ☆ Agricultural Marketing
- ☆ Rural Development
- ☆ Social Forestry
- ☆ Success Story
- ☆ Technology Transfer

Different sub-categories considered for the farm information broadcasted through Interview are:

- ☆ Agricultural Policy
- ☆ Crisis
- ☆ Farm Women
- ☆ Health and Sanitation
- ☆ Agricultural Marketing
- ☆ Rural Development
- ☆ Rural Youth
- ☆ Social Forestry
- ☆ Success Story
- ☆ Technology Transfer

Different sub-categories considered for the farm information broadcasted through Radio Play are:

- ☆ Agricultural Policy
- ☆ Farm Women

- Health and Sanitation
- Rural Development
- Rural Youth
- Social Forestry
- Success Story
- Technology Transfer

Different sub-categories considered for the farm information broadcasted through Panel Discussion are:

- Agricultural Policy
- Farm Women
- Health and Sanitation
- Agricultural Marketing
- Rural Development
- Rural Youth
- Social Forestry
- Technology Transfer

C. Reliability Test of Categories and Sub-categories of Farm Information Communicated through Radio

As it is mentioned earlier, reliability test of the process of categorization and sub-categorization is very much essential in content analysis; the same Krippendorff's Alpha (1997) has been calculated for the purpose of reliability test of the categorization and sub-categorization as it is done in case of newspapers. The value of alpha for different categories and sub-categories of farm information communicated through radio was found 0.817. By following decision rule (alpha= 0.8 or above), it is found that the process of categorization and sub-categorization were reliable.

D. Time Allotted

Like 'space provided' for farm information in newspapers, time allotted for farm information in radio is also a very important indicator to assess the relative importance given to any information by radio network. Hence, time allotted for the farm information by *All India Radio, Kolkata* has been measured in the study. Total broadcasting time of *All India Radio, Kolkata* have been calculated by summing up the total broadcasting time in every single day in a week and subsequently adding them for whole year of study period. Time allotted for farm information broadcast is measured in minutes by adding the total time allotted for farm information each day for the whole year and are categorized into different strata for calculating frequency distribution:

Time Allotted		*Assigned Code*
Up to 10 minutes	-	1
Above 10 minutes to 20 minutes	-	2
Above 20 minutes to 30 minutes	-	3
Above 30 minutes to 40 minutes	-	4
Above 40 minutes to 50 minutes	-	5

Credibility Study of Farm Information Communicated through Newspapers and Radio from Technical Point of View.

A) Selection of Sample Farm Information

200-sample farm information was selected from four newspapers through Random Sampling (Without Replacement) in the proportion of farm information published in those newspapers for credibility study.

200-sample farm information was also selected from the total farm information broadcasted through Radio by following Random Sampling Method (Without Replacement) for credibility study.

Credibility Parameters

Fifteen credibility parameters have been selected after exhaustive study of existing literature for the credibility study of farm information communicated through newspapers and radio. These are:

- ✰ Presentation
- ✰ Brevity
- ✰ Detailed
- ✰ Understandability
- ✰ Clarity
- ✰ Necessity
- ✰ Feasibility
- ✰ Universality
- ✰ Timeliness
- ✰ Current
- ✰ Repetition
- ✰ Biased
- ✰ Accuracy
- ✰ Factual
- ✰ Believability

Credibility of Farm Information communicated through newspapers and radio is measured with the help of a modified 5-point semantic scale (From Strongly Agree to

Strongly Disagree) of Gaziano and McGrath (1986). The scale is given below:

Strongly Agree	*Agree*	*Undecided*	*Disagree*	*Strongly Disagree*
5	4	3	2	1
(-) 1	(-) 2	(-) 3	(-) 4	(-) 5

Reliability Test

The scale developed by Gaziano and McGrath (1986) has been modified according to the need of the study. Hence, reliability test of this modified scale has been done on randomly selected 200-sample farm information, of which, 100- sample farm information were selected from four newspapers (taking 25 farm information form each newspapers) and 100 sample farm information from radio. Reliability of the scale is measured for the farm information communicated through both newspapers and radio by *Internal Consistency Method* by measuring *Cronbach's Alpha* (*Cronbach,* 1951). Inter-correlation matrix of the fifteen credibility variables was prepared. It was measured 0.746 and decision rule is that when Alpha value is 0.7 and above, subject under study is considered reliable. Hence, this scale used for the present study was found reliable.

Operationalization of Different Terms Used in the Study

Farm Information

For the present study, any meaning, knowledge instruction, communication, representation and mental stimulus regarding farming, farm people and farm practices published and broadcasted in newspapers and radio respectively have been considered as farm information.

News

For the present research any new farm information, which is current and related to agricultural farming published in daily newspapers, is considered as 'News'.

Views

Logical opinion, personal belief or judgment regarding farm information in newspapers is considered as 'views' for the present study.

Advertisements

Paid promotion of goods, services, companies and ideas in relation to farm and home business published in newspaper is conceived as 'advertisement' for the study.

Deliberation

The farm information, which has been broadcasted by radio in the mode of presenting farm information in straight talk in nature, is conceived as 'deliberation' in the study.

Interview

The farm information broadcasted by radio in mode of conversation between two or more people where questions are asked to obtain information about farm business have been considered as 'interview'.

Radio Play

The play or drama broadcasted by radio and disseminate farm information is considered as radio play in the present research.

Panel Discussion

The structured conversation on given topic of farming, farm people and farm business in presence of a mediator and broadcasted by radio is considered as panel discussion for the study.

Agricultural Policy

Any farm information that describes a governments set of laws relating to domestic agriculture or relates the policy matter in global scenario has been categorized as agricultural policy.

Agricultural Pollution

Different kind of pollution related farm business and agricultural operation is considered as agricultural pollution in the study.

Agricultural Crisis

Any unstable situation of extreme danger or adverse condition regarding farming, published or broadcasted is conceived as 'agricultural crisis'.

Agricultural Marketing

Farm information regarding planning and implementation strategies for buying, selling, distribution and services of agricultural inputs and produces are categorized under 'agricultural marketing' in the study.

Natural Calamities

Any disasters or natural event resulting in great loss and misfortune that negatively affects farm business temporarily or permanently is considered as 'natural calamities' for the present study.

Social Conflict

Farm information depicting a state of opposition or disagreement between two persons or group in the society and which related to farm business is conceived as 'social conflict' for this study.

Social Forestry

Farm information published related to science, art and practices of managing and using the natural resources that occurs on and in association with forest land for the human benefit in society are regarded as 'social forestry' in the study.

Success Story

The story related to success of an individual or group or organisation in the field of farm business and which can help to inspire others as a reference has been considered as 'success story' in the present study.

Technology Transfer

Farm information published or broadcasted related to any activity that lead to the adoption of new technique, theoretical and practical skills and know how of farm business from the owner of technology to users or beneficiaries are conceived as 'technology transfer' in this study.

Rural Development

The farm information, which is related to the different development activities for rural people and rural areas have been considered as 'rural development' in this study.

Weather Forecast

Any farm information communicated to assess the future state of the atmosphere with respect to precipitation, clouds, wind and temperature over geographical area of study and which help in the operation of farm business is considered as 'weather forecast' for the study.

Agricultural Machinery

The farm information related to any impact application of technology, which help in various agricultural operations have been, conceived as 'agricultural machinery' for the study.

Fertilizer

Any 'advertisement', which promotes the organic or inorganic plant food, which may be either liquid or granular, used to amend the soil in order to improve the quality or quantity or plant growth has been categorized as fertilizer in this study.

Fishery

Any advertisement, which promotes the activities related to fishery, has been designated as 'fishery' in this study.

Seed

The advertisements, which provide information, related to seed, seed material or anything related to seed have been considered as seed.

Co-operative

Any farm information regarding a duly registered and democratically controlled association of persons, with a common bond of interest, who have voluntarily joined together to achieve a lawful common social or economic end and particularly helps in farm business activities have been considered as 'co-operative' for the study.

Health and Sanitation

Farm information related to comprehensive approach for the development of public health of rural people and procedures to prevent the spread of disease has been considered as 'health and sanitation' in the study.

Rural Youth

Information broadcasted for the young, immature rural people are categorized as 'rural youth' in the study.

Farm Women

Any farm information, which is related to women, engaged in farm business, farming practice and manages the activities of farm households are considered as 'farm women'.

Credibility

For the present study 'credibility' is conceived as the trustworthiness of farm information published and or broadcasted. For ascertaining the credibility score fifteen credibility parameters are identified and selected from the past studies. Credibility of farm information have been measured in 5- point semantic scale ranging from strongly agree (±5) to strongly disagree (±1) separately for two categories of respondent, namely farmers and researcher. Cumulative score of all the fifteen credibility parameters constitute the total credibility score for the farm information under question.

Presentation

The attribute of presenting the content of farm information is measured as presentation in 5-point semantic scale, ranging from strongly agree (±5) to strongly disagree (±1). Presentation is considered as a positive trait.

Brevity

Attribute of being brief expressions of farm information is measured by brevity in a 5-point semantic scale from strongly agree (±5) to strongly disagree (±1). It is considered as positive trait.

Detailed

Degree to which farm information is developed or executed with care and in minute detail is considered as 'detailed' in the present study. It has been measured in 5-point semantic scale ranging from strongly agree (±5) to strongly disagree (±1). It is considered as positive trait.

Understandability

Degree of ability to which the language of farm information being understandable has been conceived as understandability. It has been measured in 5-point semantic scale ranging from strongly agree (±5) to strongly disagree (±1). It is considered as positive trait.

Clarity

The comprehensibility of clear expression of farm information is conceived as clarity and it has been measured in 5-point semantic scale ranging from strongly agree (±5) to strongly disagree (±1). 'Clarity' is considered as positive trait.

Necessity

The extent of degree to which farm information is essential or indispensable is considered as necessity. It has been measured in 5-point semantic scale ranging from strongly agree (±5) to strongly disagree (±1). It is considered as a positive trait.

Feasibility

'Feasibility' has been operationalized in the study as the extent to which the farm information published or broadcasted may be done practically and successfully. It has also been measured in 5-point semantic scale ranging from strongly agree (±5) to strongly disagree (±1). 'Feasibility' is considered as positive trait.

Universality

Degree to which farm information is applicable over a large geographical area and for every kind of farmer is measured with universality in 5-point semantic scale ranging from strongly agree (±5) to strongly disagree (±1). It is considered as positive trait.

Timeliness

Attribute of presenting farm information at the right time has been considered as timeliness in this study. It has been measured in 5-point semantic scale ranging from strongly agree (±5) to strongly disagree (±1). It is considered as a positive trait.

Current

Attribute of occurring in or belonging to the present time of farm information is considered as current, in the study. It has been measured in 5-point semantic scale ranging from strongly agree (±5) to strongly disagree (±1). It is considered as a positive trait.

Repetition

'Repetition' has been operationalized as the publishing or broadcasting of farm information, which has published or broadcasted before. It has also been quantified in 5-point semantic scale ranging from strongly agree (±5) to strongly disagree (±1). It is considered as a positive trait.

Biased

Attribute of being influenced in an unfair way has been considered as 'bias' in the study. It has been measured in

5-point semantic scale ranging from strongly agree (±5) to strongly disagree (±1). It is considered as a negative trait.

Accuracy

'Accuracy' has been operationalized as the extent to which farm information is truthful or valid in what it says about. It has been measured in 5-point semantic scale, ranging from strongly agree (±5) to strongly disagree (±1). It is considered as a positive trait.

Factual

Degree to which farm information is characterized by fact is considered as 'factual' in the present research. It has been measured in 5-point semantic scale, ranging from strongly agree (±5) to strongly disagree (±1). It is considered as a positive trait.

Believability

Believability has been operationalized in this study as a way to measure the quality of different farm information of being believable or trustworthy. It has been measured in 5-point semantic scale ranging from strongly agree (±5) to strongly disagree (±1). It is considered as a positive trait.

Study of Farm Information Communicated through Newspapers and Radio as Perceived by the Farmers

A. Locale of Research

To study the farmers' perception of farm information communicated through newspapers and radio, 200-sample farmers were interviewed with the structured schedule.

a) Selection of State

The State of West Bengal has been selected purposively because majority of the population in the state communicate in Bengali as it is their mother tongue and Newspapers and Radio programme selected for the study communicate farm information in Bengali. The Newspapers and radio programme are published and broadcasted from State Capital, Kolkata and cover almost every corner of the state.

b) Selection of District

The district of Birbhum has been selected purposively for the study due to easy accessibility to the researcher and all the four newspapers and radio programme are also accessible in the district.

c) Selection of Blocks and Respondents

All nineteen blocks of the district have been selected for the study. A multistage sampling procedure was followed to find the target respondent. For the present study those farmers were considered as respondent who read at least one of those selected four newspapers and follow at least one of those selected three radio programmes in regular basis. In other words respondents had to be literate. The Birbhum district consists of three Sub-divisions; Bolpur, Suri and Rampurhat comprising of 19 blocks and the total number of cultivators are 2,60,045 having a literacy percentage of 67.31 per cent of total cultivators (Census Data 2001, Govt. of India), which means 1,75,036 cultivators are literate and have the basic eligibility to read newspapers. At the first stage of sampling 1,750 literate cultivators (1 per cent of total literate cultivators in the district) were selected through simple random sampling (without replacement) method from the list of literate farmers in the district. Now, detailed address of the selected 1750 farmers have been colleted from Government Administrative Offices in the district and they have been surveyed during pilot study, whether they fulfill the above mentioned criteria to be a respondent of the study. From the pilot survey, it has been found that only 937 farmers fulfilled the basic criteria to be selected as responded which constituted the size of the population to be studied. The sub-division wise distribution of the specified population is presented below:

Sub-division	*Number of the Farmers Having Basic Criteria to be Selected as Responded*
Rampurhat	374
Suri	299
Bolpur	264
Total	937

Now, the basic premise for selection of target respondents of a size of around 20 per cent of the population as the sample, the names of 200 respondents (21.34 per cent of the total population of 937) from all the blocks for the final study have been selected by Random Sampling (Probability Proportion) Method in proportion to the numbers of eligible cultivators (937 farmers as mentioned above). Accordingly, the actual numbers of sample farmers selected from each block is given in Table 2.

B. Pilot Study

Before conducting the final study, a pilot study has been conducted on ninety respondents (thirty farmers randomly selected from each three sub-divisions) to understand the basic situational background of the respondents. It also help to standardize the structured schedule and for the reliability test of the credibility scale. This study helped to addition, alteration and elimination of some techniques and respondents as for need of the study. As for example; Television was also selected for the study, but it has been omitted in the present study due to some situational problem, which has been mentioned earlier.

Table 2: Distribution of the Respondents

Sl.No.	*Name of the Sub-Divisions/Blocks*	*Total Cultivators*	*No. of Selected Respondents (200)*
Rampurhat Sub-Div.		**103905**	**80**
1.	Murarai – I	7401	7
2.	Murarai - II	11572	9
3.	Nalhati - I	15900	12
4.	Nalhati - II	8413	6
5.	Rampurhat - I	18272	14
6.	Rampurhat - II	12233	9
7.	Mayureswar - I	15385	12
8.	Mayureswar - II	14729	11
Suri Sub-Div.		**83092**	**64**
9.	Mohammad Bazar	10836	8
10.	Sainthia	18033	14
11.	Dubrajpur	17460	13
12.	Rajnagar	7290	6
13.	Suri - I	6345	5
14.	Suri - II	10554	8
15.	Khoyrasol	12574	10
Bolpur Sub-Div.		**73048**	**56**
16.	Bolpur-Sriniketan	16387	13
17.	Labpur	20743	16
18.	Nanoor	21508	16
19.	Illambazar	14410	11

Source: Census Data 2001, Govt. of India.

C. Socio-Economic Variables

Socio-economic condition of the farmers is very important dimension in terms of accessibility of the mass media and also credibility aspects of farm information. So, different socio-economic variables have been selected for the study to assess this dimension. They are as follows:

a) Educational Status of the Farmer

Educational status of the farmers has always some impact over accessibility of the media, particularly on print media and radio as well. To quantify the educational status of the respondents, the scoring system followed by Pareek and Trivedi (1964) in their 'Socio-economic Status Scale-Rural' has been used. The scoring is as follows:

Level of Education	*Score*
Illiterate	0
Can read only	1
Can read and write	2
Primary	3
Middle School	4
High School	5
Graduate and Onwards	6

b) Family Size

Family size is another important variable to assess the socio-economic condition of the farmers. It has been measured by the 'Socio-economic Status Scale- Rural' developed by Pareek and Trivedi (1964). It is as follows:

Family Size	*Score*
Up to 5 members	1
Above 5 members	2

c) Family Type

Exhaustive study of literature also suggests that family type has also very important role on socio-economic condition of the farmers. It refers whether a farmer's family is single or joint. To quantify the family type of farmers, the scale developed by Pareek and Trivedi (1964) in their 'Socio-economic Status Scale- Rural' has been followed:

Family Type	*Score*
Single	1
Joint	2

d) Holding Size

This research is mainly concentrated on the study of farm information. In this regard, holding size of the farmer is another very important variable. In the present study, to measure the land holding status, the scale developed by Pareek and Trivedi (1964) has been used:

Land Holding	*Score*
No Land	0
Up to 1 acre	1
Up to 5 acres	2
Up to 10 acres	3
Up to 15 acres	4
Up to 20 acres	5
Above 20 acres	6

e) Social Participation

'Social participation' plays very important role in socio-economic condition of the farmers. This refers to the degree of involvement of an individual in the social event and issues. To measure this respondents have been asked whether they are member or office bearer in some organization listed in the structured schedule. The scale used in the study was developed by Pareek and Trivedi (1964).

Category	*Score*
No Member	0
Member of one organization	1
Member of more than one organization	2
Office bearer of one organization	3
Office bearer of more than one organization	4
Distinctive type (Pradhan, Sabhapati etc.)	5

f) Exposure to Channels of Farm Information Communication

It refers to exposure of an individual to different channels of farm information like Administrative Set-up, Mass Media etc. To measure the exposure to channels farm information among the respondent, they have been asked to indicate on a 4-point continuum as how often he got farm information from different channels. The scoring procedure for the responses was most often-3, often-2, sometimes-1 and never-0. The score for an individual respondent has been obtained by adding the scores over different channels.

D. Preference Orders

Preference orders by the farmers regarding farm information communication through newspapers and radio have been done to understand the need and interest of the people. Different preference orders have been studied and measured the farmers' responses through frequency distribution. They are:

- ✰ Preference Orders of Different Bengali Newspapers for Getting Farm Information
- ✰ Preference Orders of Different Farm Information Published in Newspapers
- ✰ Preference Orders of Different Radio Programmes Communicate Farm Information by AIR, Kolkata
- ✰ Preference Orders of Different Farm Information Broadcasted by AIR, Kolkata.

E. Credibility of Farm Information as Perceived by Farmers

To understand the perception of the farmers regarding credibility of farm information communicated by newspapers and radio; this credibility study has been done on 200-sample farmers. The credibility scale mentioned earlier has been applied for the purpose separately for the farm information communicated through newspapers and radio.

Like earlier cases reliability score measured for the credibility scale on 50 farmers applying Internal Consistency Method and it was measured 0.782, which is permissible.

Statistical Techniques Applied

To satisfy the different objectives and to test the hypotheses taken for the study, different statistical techniques have been applied which are mentioned below. All the statistical analyses have been done by statistical software like SPSS-11.0, MINITAB and MS-Excel Spreadsheet etc.

I. Reliability Measures

For reliability measures Krippendorff's Alpha has been used to test the reliability of categorization of farm information communicated through newspapers and radio and Internal Consistency Method (Cronbach Alpha) for reliability of credibility scales.

a) Krippendorff's Alpha

α-Agreement for Reliability of coding by Krippendorff's α (1992)

$$\alpha = 1 - \frac{D_o}{D_e}$$

where,

D_o: Measure of the observed disagreement

D_e: Measure of the expected disagreement.

b) Internal Consistency Method

Cronbach Alpha (*Cronbach,* 1951) has been used to measure reliability of credibility scale through Internal Consistency Method. It can be expressed as follows when correlation matrix is used as it has been done in the present study:

$$\alpha = \frac{N\bar{p}}{[1+\bar{p}\,(N-1)]}$$

where,

N: Number of items

$\bar{\rho}$: Mean inter-item correlation

II. Frequency Distribution

III. Mean

It is a measure of central tendency and is represented by a single figure, which describe the entire series of observation with their varying sizes. Mean of a set of observation is defined as their sum, divided by the number of observation. Mathematically it may be represented as:

$$\overline{X} = \frac{1}{N}\sum f_i X_i$$

where,

$\overline{X}$: Weighted Arithmetic mean

N: Number of observation

f: Frequency with which observation occur

i: Any number (*e.g.* 1,2,3,.......) representing the position of the particular observation

X: Value of observation

IV. Standard Deviation

Standard deviation is a measure of dispersion, which implies extent to which observation varies among them. Standard deviation of a set of observation is the square root of the arithmetic mean of squares of deviation from arithmetic mean. For frequency distribution Standard Deviation (SD) is measured as follows:

$$\sigma = \sqrt{\frac{1}{N\sum f_i\left(X_i - \overline{X}\right)}}$$

where,

σ: Standard deviation

N: Total number of observation

$\overline{X}$: Value of observation in a particular cell

f: Frequency of observation in the corresponding cell

X: Mean of the observation

i: Any number (*e.g.* 1,2,3,.......) denoting position

V. Co-efficient of Correlation

The term correlation is used to denote the degree of association between variables. In this study it is concerned with linear correlation, which is measured by correlation co-efficient. It is calculated by using following formula:

$$r_{xy} = \frac{\Sigma XY - n\overline{XY}}{\sqrt{\left[\left(\Sigma X^2 - n\overline{X}^2\right)\left(\Sigma Y^2 - n\overline{Y}^2\right)\right]}}$$

where,

X: Observed value of independent variable

Y: Observed value of dependent variable

$\overline{X}$: Mean value of the observation of the independent variable

$\overline{Y}$: Mean value of the dependent variable

n: Total number of observation

r_{xy}: Co-efficient of correlation between independent (X) and dependent variable (Y)

Test for Significance of Observed Correlation Co-efficient

For testing significance of simple correlation co-efficient (r-value), the following formula is used:

$$t = \frac{r\sqrt{(n-2)}}{\sqrt{1-r^2}}$$

where,

r: Correlation co-efficient value

n: Total number of observation

t: Test statistics

The test statistics follows student's *t* distribution with (n-2) degrees of freedom in two-tailed test.

VI. Factor Analysis

Factor is a method of multivariate analysis and is often used as method of data reduction. The different steps of factor analysis have been employed in the study are as follows:

- ☆ Correlation Matrix is used as a primary data for factor extraction.
- ☆ Principal Component Analysis is applied as a method of extraction of components.
- ☆ Variables which have Eigen value more than 1 are selected as components.
- ☆ Varimax type of rotation of Orthogonal Rotation with Kaiser Normalization is applied as a method of rotation.

VII. Paired *t*-Test

t-test is used for comparing the means of two independent groups. If there is any kind of correspondence between the individual values in the two samples, they should be paired and differences taken and analyzed directly as per formula:

$$t = \frac{\bar{d}}{S(\bar{d})}$$

where,

d: Mean of the differences in each pair

n: Pairs of observations

$S(\bar{d})$: Standard error of $(\bar{d})$ and

$$S(\bar{d}) = \frac{\sqrt{s^2}}{n}$$

Again, $$s^2 = \frac{1}{n-1}\left[\Sigma d^2 - \frac{(\Sigma d)^2}{n}\right]$$

Information about Newspapers

Aajkal is one of the leading Bengali newspapers in Kolkata, India. The headquarter is located at BP-7, Sector-V, Salt Lake, Kolkata. The newspaper arose from the transformation of the newspapers industry in the 1980s and was started in 1981. The newspaper is well known for its comprehensive sports news.

Ananda Bazar Patrika is one of the premier Bengali language daily newspaper in India. It is published from 6 & 9, Prafulla Sarkar Street, Kolkata, and simultaneously from Siliguri, and has a national circulation. The paper was founded in 1922. It is the most circulated Bengali daily.

Bartaman is the only anti-Establishment Bengali language newspaper published from Kolkata. Apart from the Kolkata edition, the newspaper has four other simultaneous editions, published daily from four major towns of West Bengal – Siliguri, Burdwan, Malda and Midnapore. The newspaper was founded in December 1984. Bartaman Pvt. Ltd., which was originally centered at 76A, AJC Bose Road, has recently shifted its entire operations to a new state-of-the-art building on 6, JBS Haldane Avenue (Eastern Metropolitan Bypass), Calcutta - 700105. The building was officially inaugurated on 07.12.2005.

আজকাল

রুনির অভাব বোঝা গেল

ভয়ের ডায়েরি

দেশের হয়ে বেকহ্যাম
লিয়েভারের মতো!

সব প্রশ্নেই এরিকসন:
৩ পয়েন্ট তো পেলাম

নামল মারাদোনার দেশ
মেসিকে বিশ্রামে রাখা হল!

JVC

বিশ্বকাপের বিশেষ মুহূর্তগুলো ধরে রাখুন
Kodak EasyShare ডিজিটাল ক্যামেরায়

Figure 1: Front Page (*Aajkal*)

আনন্দবাজার পত্রিকা

সৌরভের মন্ত্রপূত হার ঘর থেকে হাতিয়ে পাকড়াও চোর

এশিয়ার মন ভরাতে ব্যর্থ কোরিয়াও

ছন্দ আছে, ভাগ্য থাকলে কিন্তু চমকে দিত টোগো

বালাক ফিরছেন, তবু আবেগের ম্যাচে নজর আজ পোডোলস্কি-ক্লোসের দিকে

৩ হাজার কোটির লগ্নি হলদিয়ায়, সৌজন্য আইওসি

86 Aakashians passing WBJEE (Medical) '06 swear by our medicine for SUCCESS

Aakash Institute

Figure 2: Front Page (*Ananda Bazar Patrika*)

বর্তমান

বামেদের সঙ্গে বিরোধ, রাজ্যে আসছেন না প্রধানমন্ত্রী

তেলের দাম ক
চাপ সোনিয়া
আপত্তি মনমো

দক্ষিণবঙ্গে ভর্তির জন্য হুড়োহুড়ি

ভিসেরা রিপোর্টে মিলল হেরোইন ও কোকেন

চক্রান্ত করে খুন না মাদকে মৃত্যু বিবেক মৈত্রের, রহস্য বাড়ছে

পরিবহণ মন্ত্রীর কাছে আজ বাস মালিকরা

রাহুলের সঙ্গে মাদক চক্রের যোগ আছে কিনা দেখছে পুলিশ

জমি বিতর্ক: মুখ খুলতে নিষেধ করে রেজ্জাককে সতর্ক করল পার্টি

বিশ্বকাপের বাজির দরে অনেকটাই এগিয়ে ব্রাজিল

পি সি চন্দ্র জুয়েলার্স

৯ থেকে ১২ জুন, ২০০৬ (রবিবার খোলা)

LUCKY DRAW

P. C. CHANDRA JEWELLERS

Figure 3: Front Page (*Bartaman*)

প্রতিদিন

আমাদের টিম

ব্রাজিলের অনুশীলনে উচ্ছ্বাস ● রুনির স্ক্যান রিপোর্ট বিশ্লেষণ করতে ডাক্তার নিয়োগ কিফার

মায়ের অপেক্ষায় রয়েছেন রোনাল্ডিনহো

ফার্গির ভয়ে ছোঁ মেরে মার্সিডিজে

জ্বালানির পর বিলগ্নিতেও কড়া মনমোহন

কট্টরপন্থীরা ঝড় তুলবেন

মাদককাণ্ডে ধৃত প্রমোদের আর এক সচিব

পশ্চিমবঙ্গ প্রযুক্তি বিশ্ববিদ্যাল

পেশাদার স্নাতক ডিগ্রি কোর্স

জর্জ কলে

Figure 4: Front Page (*Sambad Pratidin*)

Sambad Pratidin is another widely circulated newspaper in West Bengal and it publishes from 20,Prafulla Sarkar Street, Kolkata.

Information about All India Radio

As it is mentioned earlier that *All India Radio, Kolkata* broadcast three Farm and Home Programme everyday. These programme are:

a) Chasi Bhaider Bolchhi (Addressing Farmers)

Chasi Bhaider Bolchhi is broadcasted everyday in the morning at 6.30 A.M to 6.40 P.M for ten minutes.

b) Uno Jamir Duno Phasal (Double Crop in Fertile Land)

This programme is broacasted from 12.40 P.M to 01.00 P.M for twenty minutes. This programme broadcsated with different name for different name. For easy undrstanding this particular name, *Uno Jamir Duno Phasal* has been taken in this research.

c) Krishi Kathar Asar (Programme for Agriculture)

This programme is broadcasted in the evening hours from 06.40 P.M to 07.30 P.M. but time schedule is not always same in everyday. Often more than one farm information are broadcasted in this programme.

Information about Birbhum District

To understand the farmers' perception regarding farm information communicated by newspapers and radio this study has been conducted in Birbhum district of West Bengal. Birbhum district is situated in the Rhar (literary meaning land of wild and cruel people possibly late converts to Aryan civilization but presently used as land of the red earth) region of Bengal, Birbhum was the heartland

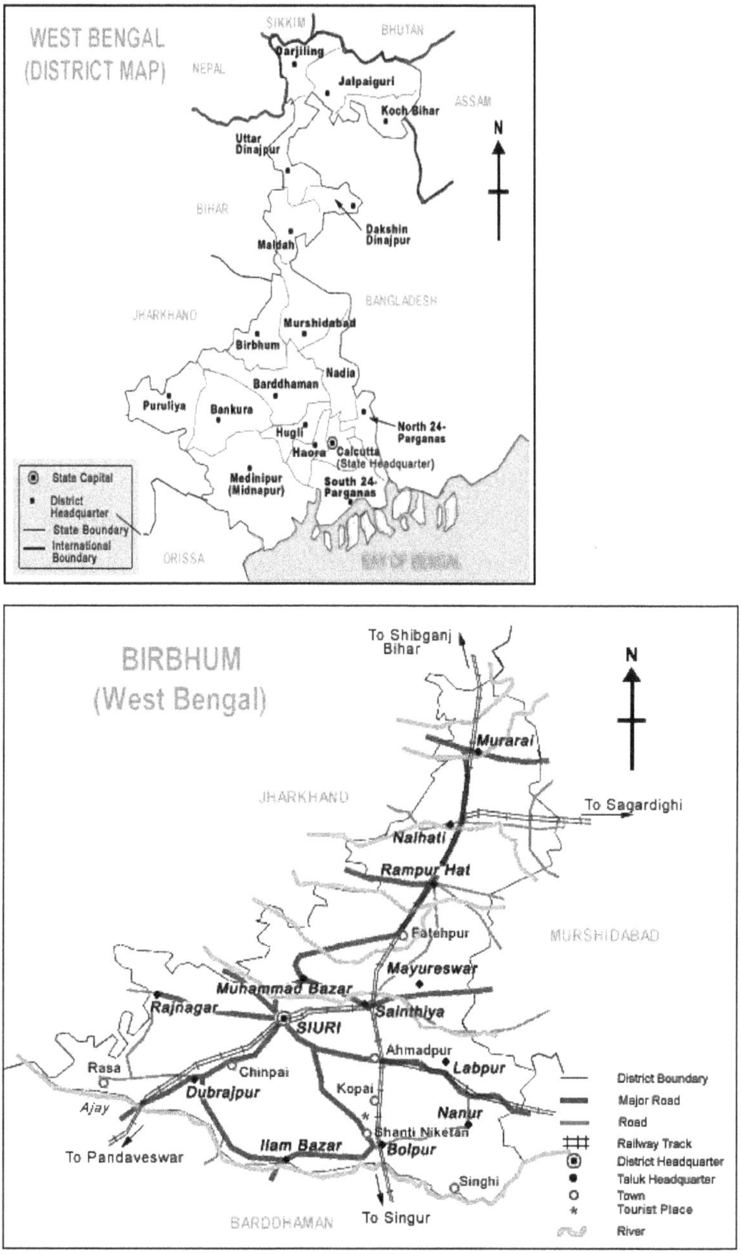

Figure 5: Geographical Maps of West Bengal and Birbhum District

of the ancient Gauda Kingdoms. It was also famous for arts and culture. Joydeva, the great Sanskrit poet and many other famous scholars lived and worked in this part of Bengal. Even to this date the Baul singers (folk singers singing with one or two string home made musical instruments) of this district maintain the rich tradition of folk art and culture. In modern times, Gurudev Rabindranath Tagore had set up the famous Visva Bharati University at Santiniketan near Bolpur in this district.

Geography

Birbhum is the northernmost District of the Burdwan District. It lies between 23° 32' 30" North to 24° 35' 0" North latitude and 88° 1' 40" East to 87° 5' 25" East longitude. In shape it looks like an isosceles triangle. The apex is situated at the northern extremity, which is not far from the south point where the river Ganges and the hills of the Santhal Parganas begin to diverge while the river Ajay forms the base of this triangle. Birbhum is bounded on the north and west by Santhal Parganas, on the east by the districts of Murshidabad and Burdwan and on the south by Burdwan and the river Ajay separates the district from Burdwan. The district extends over an area of 4545 sq.kms.

Subdivisions

- ☆ Suri
- ☆ Bolpur
- ☆ Rampurhat

Rivers

The district is well drained by a number of rivers and rivulets running in nearly every part from west to east with a slight southerly inclination. There are two main rivers, the Maurakshi and the Ajay, the latter of which marks the southern boundary, while the Maurakshi runs through Birbhum from West to East.The other rivers of Birbhum are Bakreswar, Kopai, Dvarka, Brahmani, Hinglo, Chapala, Bansoli, Pagla etc. A project on the Mayurakshi (Mor, a tributary of Ganges), that includes the Tilpara barrage (near Suri), provides irrigation for about 600,000 ac (240,000 ha).

Climate

The weather in the western side is dry, extreme, relatively milder on the eastern side. The dry plateau feature makes the land less fertile compared to the plains in east. The hot weather usually last from the middle of March to the middle of the June, the rainy season from the middle of June to the middle of October, and the cold weather from middle of October to the middle of March. As a rule, the wind is from southeast in summer and from the northwest in winter. The agriculture thus dependent on irrigation from the rivers and the rainy season.

Landscape and Vegetation

Important geographical features in Birbhum are the khoai (eroded undulated patches of barren area), the red laterite soil ("Ranga mati"), palm trees visible from long distance and multiple protected forested areas in the western side. The most interesting part of the geology of Birbhum stems from the gradient of red soil. If we look at the soil map of India we will notice that actually the Rarh region has lateritic soil (type 28 in the map), and the Vajjabhumi has a different kind of sandy hard red soil of the "Alfisol" type. Birbhum is one of the major mineral reserves of Bengal. Coal, China clay, iron-ore and granite

Table 3: Distribution of Population in Birbhum District

Name of District/ Town/C.D. Block/Out Growth/Village	*Area Vill=Hectares CD/Town= km²*	*No of House-hold*	*Population (Person)*	*Population (Male)*	*Population (Female)*	*Total No. of Villages*	*Total No. Inhabited of Villages*	*Total No. of Uninhabited Villages*
BIRBHUM	4545.00	598429	3015422	1546633	1468789	**2478**	**2260**	**218**
Murarai - I	175.51	28038	154342	79214	75128	85	80	5
Murarai - II	185.33	33229	177748	90473	87275	72	67	5
Nalhati - I	249.71	40474	208642	107527	101115	99	94	5
Nalhati - II	109.15	20627	107658	54976	52682	48	48	
Rampurhat - I	287.63	31040	159193	81349	77844	119	113	6
Rampurhat - II	181.55	30325	158742	81919	76823	95	92	3
Mayureswar - I	224.83	28376	139733	71650	68083	114	108	6
Mayureswar - II	156.57	23246	113031	58185	54846	131	125	6
Mohammad Bazar	315.64	27805	139465	71384	68081	158	138	20
Rajnagar	221.47	14470	69692	35512	34180	99	89	10
Suri - I	154.65	20047	96476	49939	46537	114	105	9
Suri - II	135.81	15870	77001	39590	37411	95	85	10
Sainthia	304.39	36014	175645	90048	85597	230	216	14

Contd...

Table 3–*Contd...*

Name of District/ Town/C.D. Block/Out Growth/Village	*Area Vill=Hectares CD/Town= km²*	*No of House-hold*	*Population (Person)*	*Population (Male)*	*Population (Female)*	*Total No. of Villages*	*Total No. Inhabited of Villages*	*Total No. of Uninhabited Villages*
Labpur	267.98	35450	176865	91065	85800	180	161	19
Nanoor	311.83	38589	193775	98801	94974	138	132	6
Bolpur Sriniketan	334.58	37280	178111	90883	87228	170	159	11
Illambazar	261.54	29015	145182	74728	70454	134	128	6
Dubrajpur	344.88	31615	159011	81369	77642	227	194	33
Khoyrasol	272.19	25265	135101	69785	65316	170	126	44

Table 4: Educational Facilities in the Birbhum District

7	No. of Villages having Primary School within the Village	Distance Range of Nearest Primary School If Not Available within the Village			No. of Villages having Middle School within the Village	Distance Range of Nearest Middle School If Not Available within the Village			No. of Vill. having Sec. School within the Village	No. of Vill. having Sr. Sec. School within the Village	No. of Vill. having Colleges within the Village	Distance Range of Nearest College If Not Available within the Village		
		< 5 km.	5-10 Km.	10+ km.		< 5 km.	5-10 Km.	10+ km.				< 5 km.	5-10 km.	10+ km.
	2001		2001		2001		2001		2001	2001	2001		2001	
WEST BENGAL	**28185**	**8369**	**1308**	**94**	**5541**	**18628**	**12649**	**1138**	**3411**	**999**	**138**	**2894**	**8172**	**26752**
BIRBHUM	**1770**	**403**	**84**	**3**	**334**	**1061**	**835**	**30**	**222**	**44**	**10**	**125**	**603**	**1522**
Murarai - I	68	9	3	0	13	28	35	4	9	2	1	13	32	34
Murarai - II	64	3	0	0	12	39	16	0	8	1	0	2	16	49
Nalhati - I	87	5	2	0	22	34	37	1	16	4	3	11	34	46
Nalhati - II	42	5	0	1	11	31	6	0	8	2	0	0	1	47
Rampurhat - I	98	15	0	0	19	44	46	4	8	2	0	15	64	34
Rampurhat - II	81	10	1	0	22	46	18	6	16	2	0	5	8	79
Mayureswar - I	92	15	1	0	23	80	5	0	15	1	0	0	24	84
Mayureswar - II	73	41	11	0	14	75	35	1	10	3	0	0	12	113

Contd...

Table 4–*Contd...*

7	No. of Villages having Primary School within the Village	Distance Range of Nearest Primary School If Not Available within the Village			No. of Villages having Middle School within the Village	Distance Range of Nearest Middle School If Not Available within the Village			No. of Vill. having Sec. School within the Village	No. of Vill. having Sr. Sec. School within the Village	No. of Vill. having Colleges within the Village	Distance Range of Nearest College If Not Available within the Village		
		< 5 km.	5-10 Km.	10+ km.		< 5 km.	5-10 Km.	10+ km.				< 5 km.	5-10 km.	10+ km.
	2001	2001			2001	2001			2001	2001	2001	2001		
Mohammad Bazar	94	44	0	0	16	57	64	1	14	2	1	1	20	116
Rajnagar	69	9	11	0	11	42	35	1	6	2	0	0	2	87
Suri - I	75	24	5	1	16	40	49	0	12	2	1	1	75	28
Suri - II	67	18	0	0	9	56	20	0	8	2	0	12	29	44
Sainthia	155	14	47	0	22	40	152	2	18	1	0	5	62	149
Labpur	130	31	0	0	24	76	55	6	16	5	1	8	69	83
Nanoor	118	13	0	1	26	88	18	0	16	5	1	12	38	81
Bolpur Sriniketan	121	35	3	0	26	84	48	1	12	3	0	13	43	103
Illambazar	109	19	0	0	15	80	33	0	6	2	0	1	3	124
Dubrajpur	124	70	0	0	19	61	112	2	11	2	1	18	49	126
Khoyrasol	103	23	0	0	14	60	51	1	13	1	1	8	22	95

are mined in the west Birbhum. The laterite soil of western highlands is iron rich and acidic in nature. It grows characteristic shrubs, herbs and some trees. These include species of *Wendlandia, Evolvulus, Stipa, Tragus, Perotis, Spermacoce, Zizyphus, Capparis*. Trees like mango and bamboo are also frequently found. Other abundant species are jack, arjun, sal, date plum, piar, dhau, kend and mahua. Rice, legumes, wheat, corn (maize), and mangoes are the chief crops harvested largely in the east Birbhum.

Transport

The major routes of transportation are train, bus and bullock carts. Historically Palki was another common transport now seen very rarely. The old Howrah-Sahebganj loop of the eastern railway passes through this district, which had a junction at Nalhati for the connection to Ajimganj, Murshidabad district.

(Source: http: //www.hindubooks.org/temples/ bengal/birbhum/index.htm)

Chapter 4

Results and Discussion

Results of the study are presented in this chapter with the help of proper statistical analyses. Results are presented in this chapter in different segments to satisfy the different objectives taken for the study. They are as follows:

Content Analysis of Farm Information Communicated through Selected Newspapers

Farm Information Communicated through Newspapers

Total numbers of farm information consisting of all three major categories, namely; news, views and advertisements communicated through selected newspapers during the one year study period, have been computed and presented in the Table 5.

Table 5 suggests that *Bartaman* published highest number of farm information (633) during the study period among all newspapers taken for the study, whereas *Aajkal* ranked the lowest. The other two newspapers; *Ananda Bazar Patrika* and *Sambad Pratidin* were placed at the middle of the scale with 372 and 325 numbers of published farm information respectively.

Table 5: Number of Farm Information Communicated through Newspapers

Name of the Mass Media	*No. of Farm Information*
Aajkal	146
Ananda Bazar Patrika	372
Bartaman	633
Sambad Pratidin	325

Relative Importance Given to Farm Information in Selected Newspapers

No mass media covers single type of information. Newspapers and Radio, being mass media are no exception. Moreover, mass media always consider the question of 'extent of acceptance' among the mass regarding types of information. All such considerations of mass media regulate the importance to be given to different kinds or types of information. This differential importance and preferences by the media authorities are manifested in selection of farm information in respect to provision of space for farm information in newspapers. Space provided for farm information in selected newspapers has been presented in the Table 6.

Table 6 envisaged that in terms of total space covered by farm information, *Bartaman* ranked the highest, whereas, *Aajkal* ranked lowest and in terms of percentage distribution of space covered by farm information in respect to total space published by newspapers during the study period, *Bartaman* was much higher (1.01 per cent) than that of any other newspapers. In *Ananda Bazar Patrika* and *Sambad Pratidin*, spaces covered by farm information were 0.64 per cent each and in *Aajkal* it was only 0.21 per cent. Now, besides business, publishing or broadcasting farm information calls for sharing of social responsibility by the mass media authorities and it is found that among the newspapers, *Bartaman* shouldered it most.

Table 6: Space Provided for Farm Information in Selected Newspapers

Newspapers	*Total Space Published (Sq. cm.)*	*Space Covered or Time Given for Farm Information (Sq. cm.)*	*Per cent*
Aajkal	5395698	11287.82	0.21
Ananda Bazar Patrika	8264448	52784.88	0.64
Bartaman	7027228	70935.62	1.01
Sambad Pratidin	6293430	40434.02	0.64

Preferential Selection of Different Categories and Sub-Categories of Farm Information in Selected Newspapers

For a long year of critical study of all four selected newspapers, it was found that each newspaper had preferential selection for publication of different categories of farm information. In the Table 7 distribution of different categories of farm information are presented.

From the Table 7, it is found that 'News' category of farm information were preferred most by all the newspapers followed by 'Views' and 'Advertisements'. In *Aajkal*, 'News' occupied 84.25 per cent, in *Ananda Bazar Patrika* it was 71.50 per cent, in *Sambad Pratidin*, it was 69.54 per cent. For 'Views' and 'Advertisements' categories much less preference was given by all three newspapers except *Bartaman*. In *Bartaman* 'News' (47.71 per cent), 'Views'

Table 7: Distribution of Different Categories of Farm Information Published in Selected Newspapers

Category of Farm Information	*Aajkal*		*Ananda Bazar Patrika*		*Bartaman*		*Sambad Pratidin*	
	Frequency	*Per cent*	*Frequency*	*Per cent*	*Frequency*	*Per cent*	*Frequency*	*Per cent*
News	123	84.25	266	71.50	302	47.71	226	69.54
Views	15	10.27	53	14.25	253	39.97	67	20.62
Advertisement	8	5.48	53	14.25	78	12.32	32	9.85
Total	146	100	372	100	633	100	325	100

(39.97 per cent) and 'Advertisements' (12.32 per cent) were published in more balanced way.

The Table 8 depicts the distributions of different sub categories of 'News' type of farm information in the selected four newspapers. It can be observed from the table that mostly covered sub-categories for all four selected newspapers were 'Agricultural Policy', 'Crisis' and Agricultural Marketing'. They cumulatively occupied 69.1 per cent (32.52 per cent +17.07 per cent +19.51 per cent) in case of *Aajakl,* 65.03 per cent (31.58 per cent +23.68 per cent +9.77 per cent) in case of *Ananda Bazar Patrika,* 58.94 per cent (21.52 per cent +22.85 per cent +14.57 per cent) incase of *Bartaman* and for *Sambad Pratidin,* this figure was 66.37 per cent (26.11 per cent +25.66 per cent +14.60 per cent) so far as the number of farm information published during the study year in all four selected newspapers related with different sub-categories under 'News' category. It can further be observed that all the newspapers gave cumulatively same level of importance to these three sub-categories of 'News' category.

The next important sub-category under 'News' category was found 'Weather Forecast' in case of all three selected newspapers except *Ananda Bazar Patrika.* In case of *Ananda Bazar Patrika,* the second important sub-category was 'Rural Development'.

It can be stated that in terms of selection of different sub-categories under 'News' category of farm information by the selected newspapers, all four selected newspapers were found homogeneous in general with minor exceptions.

The Table 9 gives the picture of different sub-categories under 'Views'. This also has similarity with different

Table 8: Distribution of Different Sub-Categories of 'News' Published in Selected Newspapers

Sub-Category	*Newspapers*							
	Aajkal		*Ananda Bazar Patrika*		*Bartaman*		*Sambad Pratidin*	
	Frequency	*Per cent*	*Frequency*	*Per cent*	*Frequency*	*Per cent*	*Frequency*	*Per cent*
Agricultural Pollution	1	0.81	3	1.13	4	1.32	2	0.88
Agricultural Policy	40	32.52	84	31.58	65	21.52	59	26.11
Crisis	21	17.07	63	23.68	69	22.85	58	25.66
Agricultural Marketing	24	19.51	26	9.77	44	14.57	33	14.60
Natural Calamities	2	1.63	7	2.63	5	1.66	10	4.42
Rural Development	8	6.50	33	12.41	31	10.26	12	5.31
Social Conflict	1	0.81	15	5.64	9	2.98	13	5.75
Social Forestry	2	1.63	4	1.50	10	3.31	5	2.21
Success Story	10	8.13	5	1.88	5	1.66	10	4.42
Technology Transfer	3	2.44	17	6.39	10	3.31	6	2.65
Weather Forecast	11	8.94	7	2.63	50	16.56	18	7.96
Total	123	100	266	100	302	100	226	100

Table 9: Distribution of Different Sub-Categories under 'Views' Published in Selected Newspapers

Sub-Category	*Newspapers*							
	Aajkal		*Ananda Bazar Patrika*		*Bartaman*		*Sambad Pratidin*	
	Frequency	*Per cent*	*Frequency*	*Per cent*	*Frequency*	*Per cent*	*Frequency*	*Per cent*
Agricultural Pollution	1	6.67	2	3.77	1	0.40	2	2.99
Agricultural Policy	3	20.00	18	33.96	23	9.09	16	23.88
Crisis	5	33.33	10	18.87	54	21.34	18	26.87
Agricultural Marketing	3	20.00	5	9.43	6	2.37	3	4.48
Natural Calamities	0	0	0	0	2	0.79	0	0
Rural Development	0	0	6	11.32	3	1.19	3	4.48
Social Conflict	0	0	2	3.77	0	0	0	0
Social Forestry	0	0	0	0	3	1.19	0	0
Success Story	0	0	3	5.66	1	0.40	1	1.49
Technology Transfer	2	13.33	6	11.32	159	62.85	23	34.33
Weather Forecast	1	6.67	1	1.89	1	0.40	1	1.49
Total	15	100	53	100	253	100	67	100

important sub-categories under 'News' category in terms of selection of different sub-categories for publication of 'Views' type of farm information.

In case of distribution of sub categories of 'Views' type of farm information in *Aajkal*, 'Crisis' (33.33 per cent) ranked the highest, as it is observable in the Table 9. 'Agricultural Policy' (20 per cent) and 'Technology Transfer' (13.33 per cent) were the other major sub categories of farm information under views as it is found in the table. Table also shows that 'Agricultural Policy' (33.96 per cent) and 'Crisis' (18.87 per cent) were also given much more importance in *Ananda Bazar Patrika* under 'Views' category of farm information. 'Rural Development' (11.32 per cent) and 'Technology Transfer' (11.32 per cent) were also given importance under this category in *Ananda Bazar Patrika*. In case of *Bartaman*, Table 9 confirms that 'Technology Transfer' (62.85 per cent) sub-category of farm information under 'Views' category was given much more importance than that of any sub category. Other than 'Technology Transfer', 'Crisis' (21.34 per cent) was also given importance in *Bartaman*. 'Technology Transfer' (34.33 per cent) type of farm information under views category in *Sambad Pratidin* were published in greater proportion. 'Crisis' (26.87 per cent) and 'Agricultural Policy' (23.88 per cent) came next to 'Technology Transfer' as it is found in the study in case of *Sambad Pratidin*.

So, under 'Views' category; Technology Transfer, Crisis and Agricultural sub-categories were given most importance. If five important sub-categories under 'Views' are considered, no major differences were found in all four selected newspapers.

In advertisement category, *Aajkal* published most of the Co-operative (87.5 per cent) related 'Advertisements', while only one advertisement related with Social Forestry (12.5 per cent) sub category, as it is evident from the Table 10. In *Ananda Bazar Patrika,* 33.96 per cent of the 'Advertisements' came under 'Agricultural Machinery' sub-category. 'Fertilizer' (28.30 per cent) and 'Rural Development' (24.53 per cent) sub categories of 'Advertisements' were also published in *Ananda Bazar Patrika* in considerable amount as it is found in the study. Only 13.21 per cent of the advertisement was also related to 'Seed'. It is evident from the Table 10 that 'Fertilizer' (44.87 per cent) and 'Agricultural Machinery' (37.15 per cent) were the two most important subcategories of 'Advertisements' published in *Bartaman*. 12.82 per cent of 'Advertisements' also came from 'Fishery'. In *Sambad Pratidin,* majority of the 'Advertisements' were from 'Plant Protection' (31.25 per cent) as it is found in the study. 'Rural Development' (21.88 per cent), 'Agricultural Machinery' (15.63 per cent) sub-categories of 'Advertisements' were also given importance in *Sambad Pratidin*. So, no as such relationship is found among the four newspapers in terms of publishing advertisements in all the four newspapers.

Finally, it is observed that so far as the preferential selection of different sub-categories under 'Advertisements', different newspapers gave differential importance with a common tilt to 'Agricultural Machinery' and 'Fertilizer'.

Monthly Distribution of Farm Information Published in Different Newspapers

Agriculture being the seasonal activity, requirement for different kind of farm information among the farmers varies

Table 10: Distribution of Different Sub-Categories of 'Advertisements' Published in Selected Newspapers

Sub-Category	*Newspapers*							
	Aajkal		*Ananda Bazar Patrika*		*Bartaman*		*Sambad Pratidin*	
	Frequency	*Per cent*	*Frequency*	*Per cent*	*Frequency*	*Per cent*	*Frequency*	*Per cent*
Cooperative	7	87.5	0	0	0	0	5	15.63
Social Forestry	1	12.5	0	0	0	0	0	0
Agricultural Machinery	0	0	18	33.96	29	37.18	5	15.63
Fertilizer	0	0	15	28.30	35	44.87	3	9.38
Rural Development	0	0	13	24.53	1	1.28	7	21.88
Seed	0	0	7	13.21	2	2.56	2	6.25
Fishery	0	0	0	0	10	12.82	0	0
Plant Protection Material	0	0	0	0	1	1.28	10	31.25
Total	8	100	53	100	78	100	32	100

seasonally. Table 11 depicts the monthly distribution of farm information published in selected newspapers.

In case of *Aajkal*, seasonality observed in case of 'News' type of farm information. This is found that during the month of May, June and July, most of the farm information were published as well as November, December and January were found with least number of farm information. As very few numbers of 'Views' and 'Advertisements' were published during the study year in *Aajkal*, no further analysis could be made, although the month of November was found important for both categories of farm information.

In case of *Ananda Bazar Patrika*, a balance could be found in case of number of publication of 'News' and 'Views' categories of farm information. But in case of 'Advertisements', July and February had the highest number of publication.

Although almost a balance number of publications were found in 'News' and 'Views' categories of farm information published in *Bartaman*, still the range is high. A sustained importance was given to publication of farm information under 'News' category from December to March. And in *Bartaman*, higher concentration of publication of 'Advertisements' was found during the months of November to February.

In case of *Sambad Pratidin*, almost a balance picture can be observed in terms of monthly distribution of publication of farm information under 'News', 'Views' and 'Advertisements' categories. For 'News' and 'Views' categories, hardly any seasonality could be observed.

Table 11: Month-Wise Distribution of Different Categories of Farm Information in Selected Newspapers

Month	*Aajkal*			*Ananda Bazar Patrika*			*Bartaman*			*Sambad Pratidin*		
	News	*Views*	*Advertisement*	*News*	*Views*	*Advertisement*	*News*	*Views*	*Advertisement*	*News*	*Views*	*Advertisement*
April	9	0	0	18	5	2	17	16	0	15	7	2
May	12	2	0	22	4	0	15	17	0	17	9	4
June	23	1	0	24	1	1	22	19	1	22	5	2
July	13	0	1	21	5	12	23	29	7	23	5	4
Aug	10	0	0	20	4	1	32	30	4	19	4	1
Sept	11	0	0	22	4	1	12	16	1	18	3	2
Oct	11	2	0	26	2	1	35	14	3	25	4	0
Nov	5	6	7	16	3	7	10	23	12	15	8	7
Dec	5	2	0	20	5	8	38	23	19	18	6	0
Jan	5	1	0	20	6	6	32	22	10	21	2	4
Feb	12	1	0	36	9	12	25	31	14	16	3	6
Mar	7	0	0	21	5	2	41	13	7	17	11	0
Total	123	15	8	266	53	53	302	253	78	226	67	32
Average	10.25	1.25	0.67	22.17	4.42	4.42	25.17	21.08	6.5	18.83	5.58	2.67

Considering the average monthly publication of 'News', 'Views' and 'Advertisements' categories of farm information, *Bartaman* ranked first and *Aajakl* ranked last with *Ananda Bazar Patrika* at second (ranked second in case of 'News' and 'Advertisements' and ranked third in case of 'Views') and *Sambad Pratidin* at third position (ranked second in case of 'Views' and ranked third in case of 'News' and 'Advertisements').

From the Table 11, no seasonal variation in publication of farm information could be observed. Newspapers houses generally published and averaged number of farm information every month without any extra attention during any particular cropping season. The only exception was found in case of *Bartaman*, which published most number of farm information under all three categories, that are 'News', 'Views' and 'Advertisements' during the months of December to March which is considered to be *Boro* season of cultivation.

Selection of Font Size of the Headlines for Publication of Farm Information in Selected Newspapers

For publishing, different font (type face and size of letter) and different style of headlines are used for catching the eyes of the readers. Newspapers are no exception. These differential uses of font size in headlines show the importance on particular farm information by the media. In order to enumerate the importance given by the media house, font size used in the headlines for publishing farm information are measured, tabulated presented below separately for all four selected newspapers.

Table 12: Differential Use of Font Size in Headlines of Farm Information in Selected Newspapers

Font Size	*Aajkal*		*Ananda Bazar Patrika*		*Bartaman*		*Sambad Pratidin*	
	Frequency	*Per cent*	*Frequency*	*Per cent*	*Frequency*	*Per cent*	*Frequency*	*Per cent*
11 to 20	0	0	3	0.81	0	0.00	4	1.23
21 to 30	15	10.27	40	10.75	101	15.96	34	10.46
31 to 40	79	54.11	105	28.23	257	40.60	77	23.69
41 to 50	39	26.71	136	36.56	198	31.28	145	44.62
51 to 60	5	3.42	50	13.44	35	5.53	43	13.23
61 to 70	4	2.74	31	8.33	30	4.74	14	4.31
71 to 80	2	1.37	3	0.81	5	0.79	6	1.85
81 to 90	0	0.00	1	0.27	0	0.00	1	0.31
91 and Above	2	1.37	3	0.81	7	1.11	1	0.31
Total	146	100	372	100	633	100	325	100

Table 12 depicts that majority of the headlines of farm information came under 31-40 font size category in *Aajkal* (54.11 per cent) and *Bartaman* (40.60 per cent). Headlines of considerable numbers of farm information also came under 41-50 and 21-30 font size categories in both *Aajkal* and *Bartaman*. While majority of the headlines of farm information came under 41-50-font size category in *Ananda Bazar Patrika* (36.56 per cent) and *Sambad Pratidin* (44.62 per cent). At the same time, headlines of considerable proportion of farm information were published under 31-40, 51-60 and also 21-30 font size categories. It is found from the Table 12 that in case of publishing headlines of the farm information, all the four selected newspapers primarily used font size categories of 31 to 40 and 41 to 50. In that respect no significant differences was found among the selected newspapers.

Provision of Space in Newspapers for Publishing Farm Information

Provision of space for publishing any type of 'News', 'Views' and 'Advertisements' is also an index through which it can be understood well the extent of importance given to that particular 'News', 'Views' and 'Advertisements' by the publishing newspapers. For farm information also different newspapers provide different amount of space for publication. In order to assess differential importance given by different newspapers for publishing farm information in terms of providing space in newspapers, printed area under farm information for all four selected newspapers are measured and presented below.

Table 13 depicts that majority of farm information in *Aajkal* have been published under 'up to 50 sq cm' (59) and

Table 13: Distribution of Space Provided for Publication of Farm Information in Selected Newspapers

Space Covered	Aajkal		Ananda Bazar Patrika		Bartaman		Sambad Pratidin	
	Frequency	Cumulative Percentage	Frequency	Per cent	Frequency	Per cent	Frequency	Per cent
Up to 50 sq. cm.	59	40.41	68	18.28	144	22.75	60	18.46
Above 50 sq. cm. to 100 sq.cm	53	76.71	103	45.97	191	52.92	138	60.92
Above 100 sq. cm. to 150 sq.cm	19	89.72	116	77.15	180	81.36	80	85.54
Above 150 sq. cm. to 200 sq.cm	12	97.94	42	88.44	56	90.21	26	93.54
Above 200 sq. cm. to 250 sq.cm	0	97.94	5	89.78	14	92.42	5	95.08
Above 250 sq. cm. to 300 sq.cm	1	98.63	17	94.35	27	96.69	1	95.39
Above 300 sq. cm. to 350 sq.cm	0	98.63	6	95.96	7	97.8	5	96.93
Above 350 sq. cm. to 400 sq.cm	0	98.63	5	97.3	6	98.75	1	97.24
Above 400 sq. cm. to 450 sq.cm	2	100.00	2	97.84	2	99.07	2	97.86
Above 450 sq. cm.	0	100.00	8	100.00	6	100.00	7	100.00
Total	146	100.00	372	100.00	633	100.00	325	100.00

'above 50 sq. cm to 100 sq. cm' (53) space categories. A considerable proportion of farm information had also been published under 'Above 100 sq. cm to 150 sq. cm' space categories. But in case of *Ananda Bazar Patrika,* 'Above 100 sq. cm to 150 sq. cm' category ranked highest (116) and it is followed by 'Above 50 sq. cm to 100 sq. cm' (103) and 'Up to 50 sq. cm' (68) categories. In *Bartaman,* under 'Above 50 sq. cm to 100 sq. cm' categories, 191 numbers of farm information were published. 'Above 100 sq. cm to 150 sq. cm' (180) and 'Up to 50 sq. cm' (144) categories came next. In *Sambad Pratidin,* the order is same as *Bartaman,* 'Above 50 sq. cm to 100 sq. cm.' 'Above 100 sq. cm to 150 sq. cm' and 'Up to 50 sq. cm' covered most of the farm information. Cumulative percentage shows that in all the four newspapers first three categories; 'up to 50 sq cm', 'above 50 sq. cm to 100 sq. cm' and 'Above 100 sq. cm to 150 sq. cm' comprised bulk of the farm information published. Hence, it can be stated that there is no significant difference in the importance given by selected newspapers for publication of farm information in terms of space provided.

Readability of Subject Matters Published as Farm Information in Selected Newspapers

'Readability' is another aspect of any printed material, which determines its quality. 'Readability' is the index used to determine how far a printed material is readable to its reader. To measure readability, different approaches can be adopted. Here, in this study Fog Index has been calculated for each farm information published in those selected newspapers during the study period. Then mean Fog Indexes have been computed which has been presented in the Table 14.

Table 14: Fog Index of Selected Newspapers

Newspapers	*Mean Fog Indexes*
Aajkal	6.23
Ananda Bazar Patrika	6.52
Bartaman	5.98
Sambad Pratidin	5.87

Table 14 indicates that *Ananda Bazar Patrika* and *Aajkal* ranked high in terms of mean fog index. They are 6.52 and 6.23 respectively. Whereas 5.98 and 5.87 mean fog indexes are found in case of *Bartaman* and *Sambad Pratidin*. Higher fog index denotes relatively less readable subject matter. According to Koyal (1975), Fog Index score from 5.40 to 6.09 is standard and from 6.10 to 6.40 is meant for quality magazine and demands for higher educational background of the reader. In this regard, *Sambad Pratidin* and *Bartaman* were appeared to be standard newspapers to the farmers in terms of readability of the farm information. Whereas, *Aajkal* and *Ananda Bazar Patrika* having fog index of 6.23 and 6.52 demand for higher educational background. Although, literacy percentage is creeping up in India, there are many neo literate and illiterate among farming communities, farm information should be communicated in such a manner which is having higher readability, will have the better chance for acceptance as channel of farm information among the farming community.

Content Analysis of Farm Information Communicated through Radio

For the purpose of content analysis of the radio programmes, farm information broadcasted by three programmes of *All India Radio, Kolkata*, namely; *Chasi Bhaider Bolchhi, Uno Jamir Duno Phasal* and *Krishi Kathar Aasar* have been studied carefully during the study period. The Table 15 depicts the programme wise distribution of number of farm information and time allotment for farm information in different programmes.

In case of radio programmes some times more than one farm information were broadcasted in a single programme. So total number of farm information were taken and computed for frequency distribution. From the table, it is found that Radio broadcasted much more farm information (1110) than that of any newspapers. The total numbers of farm information were almost evenly distributed among the three Farm and Home broadcasts of *All India Radio, Kolkata.* Table also shows that Radio broadcasted 20235 minutes of farm information during study period. But in case of Radio, percentage distribution of time given for farm information in respect to total broadcasting time in Radio during the study period was much higher (7.09 per cent) than that of any newspapers as it is found in the study. Among the three programme broadcasted by *All India Radio, Kolkata, Chasi Bhaider Bolchhi* and *Uno Jamir Duno Phasal* were broadcasted for ten minutes and twenty minutes everyday respectively. Whereas, Krishi *Kathar Aasar* was broadcasted everyday for thirty minutes or more. This is quite manifested in their percentage distribution in terms of time allotment during the study period. That is why *Krishi Kathar Aasar* occupied the highest percentage (3.26 per cent) and *Chasi Bhaider Bolchhi* ranked the lowest (1.27 per cent) in the Table 15.

Table 15: Number and Time Allotted for Farm Information in Radio

Name of the Radio Programme	*Total No. of Farm Information*	*Per cent*	*Total Broadcasting Time*	*Time Allotted for Farm Information*	*Per cent*
Chasi Bhaider Bolchhi	362	32.61	285480 minutes	3620 minutes	1.27
Uno Jamir Duno Phasal	373	33.60	285480 minutes	7295 minutes	2.56
Krishi Kathar Aasar	375	33.79	285480 minutes	9320 minutes	3.26
Radio (Total)	1110	100.00	285480 minutes	20235 minutes	7.09

Preferential Selection of Different Categories and Sub-Categories of Farm Information in Selected Radio Programmes

Like newspapers, from one year long critical study of all three farm broadcasts of *All India Radio, Kolkata*, it was found that *All India Radio, Kolkata* had preferential selection for mode of presentation of farm information. According to the mode of presentation, four different categories have been identified, namely; Deliberation, Interview, Panel Discussion and Radio Play under which all the farm information have been categorized. Total numbers of farm information under these broad categories have been computed and are presented in the following tables. As because, all the farm information broadcasted in the morning farm programme, *Chasi Bhaider Bolchhi* by *All India Radio, Kolkata;* were deliberation in nature, so, tabulation has not been done by programme wise. Total farm information of all three farm programmes were taken together and presented under different categories and sub-categories.

Table 16: Distribution Categories of Farm Information in Radio

Category	*Frequency*	*Per cent*
Deliberation	656	59.10
Interview	197	17.75
Panel Discussion	73	6.58
Radio Play	184	16.58
Total	1110	100

In radio, during the study period total number of farm information broadcasted was 1110 as it is mentioned earlier in Table 15. The farm information broadcasted in radio have been categorized in to four different categories according to the mode of presentation. Among them 'Deliberation' ranked highest (59.10 per cent). The next categories close to 'Deliberation' were 'Interview' (17.75 per cent) and 'Radio Play' (16.58 per cent). Farm information broadcasted through Panel Discussion was only 6.58 per cent as the Table 16 suggests. So, in general, table suggests that broadcast house preferred mostly 'Deliberation' mode of broadcasting for the dissemination of farm information.

For detailed study on preferential selection, each category has been divided into different sub-categories. Farm information broadcasted in relation with different sub-categories were counted separately for each category and presented in the following tables:

Table 17: Distribution of Sub-Categories under Deliberation in Radio

Sub-category	*Frequency*	*Per cent*
Agricultural Policy	10	1.52
Crisis	2	0.30
Farm Women	8	1.22
Health and Sanitation	12	1.83
Agricultural Marketing	23	3.51
Rural Development	171	26.07
Rural Youth	1	0.15
Social Forestry	10	1.52
Success Story	3	0.46
Technology Transfer	416	63.41
Total	656	100

This table depicts the distribution of different sub-categories under 'Deliberation' category of farm information broadcasted by radio. It is found that majority of the farm information under 'Deliberation' categories were related with 'Technology Transfer' (63.41 per cent). The other major sub categories came after 'Technology Transfer' were 'Rural Development' (26.07 per cent) followed by 'Agricultural Marketing' (3.51 per cent).

Table 18: Distribution of Sub-Categories under Interview in Radio

Sub-category	*Frequency*	*Per cent*
Agricultural Policy	15	7.61
Crisis	4	2.03
Farm Women	18	9.14
Health and Sanitation	38	19.29
Agricultural Marketing	8	4.06
Rural Development	26	13.20
Rural Youth	9	4.57
Social Forestry	6	3.05
Success Story	18	9.14
Technology Transfer	55	27.92
Total	197	100

In case of 'Interview' category', the table again suggests that 'Technology Transfer' (27.92 per cent) type of farm information was broadcasted in highest number. 'Health and Sanitation' (19.29 per cent) were also given importance under 'Interview' category in Radio. Followed by 'Rural Development' (13.20 per cent).

Table 19: Distribution of Sub-Categories under Panel Discussion in Radio

Sub-category	*Frequency*	*Per cent*
Agricultural Policy	6	8.22
Farm Women	17	23.29
Health and Sanitation	3	4.11
Agricultural Marketing	3	4.11
Rural Development	7	9.59
Rural Youth	24	32.88
Social Forestry	1	1.37
Technology Transfer	12	16.44
Total	73	100

In 'Panel Discussion' (Table 19), 'Rural Youth' (32.88 per cent) and 'Farm Women' (23.29 per cent) sub-categories of farm information were given prime importance in Radio. A considerable amount of farm information was also broadcasted in Radio about Technology Transfer (16.44 per cent) under Panel Discussion as it is found from the results.

Table 20: Distribution of Sub-Categories under Radio Play in Radio

Sub-category	*Frequency*	*Per cent*
Farm Women	4	2.17
Rural Development	110	59.78
Rural Youth	1	0.54
Social Forestry	1	0.54
Success Story	37	20.11
Technology Transfer	31	16.85
Total	184	100

The Table 20 ascertains that majority of the farm information under 'Radio Play' were broadcasted on 'Rural Development' (59.78 per cent) activities. 20.11per cent of farm information were also about different 'Success Stories' (20.11 per cent) related to farm business followed by 16.85 per cent of the farm information related with 'Technology Transfer' subcategories.

Table 21: Monthly Distribution of Different Categories of Farm Information in Radio

Month	*Category of Farm Information*				
	Deliberation	*Interview*	*Panel Discussion*	*Radio Play*	*Total Farm Information*
April	32	21	4	14	71
May	54	29	4	25	112
June	53	15	6	14	88
July	68	17	12	10	107
Aug	57	16	10	13	96
Sept	63	13	11	11	98
Oct	47	12	7	12	78
Nov	54	20	4	12	90
Dec	55	18	1	18	92
Jan	59	9	5	23	96
Feb	54	11	5	16	86
March	60	16	4	16	96
Total	656	197	73	184	1110
Average	54.67	16.42	6.08	15.33	92.5

As a whole, so far as the preferential selection of different major categories and sub-categories of farm information broadcasted by *All India Radio, Kolkata* are concerned, it is found that 'Deliberation', a major category was preferred most followed by 'Interview'. In case of different sub-categories, it is found that 'Technology Transfer' and 'Rural Development' were preferred most.

Monthly Distribution of Farm Information Broadcasted through Radio

As it is mentioned earlier agriculture being the seasonal activity, requirement for different kind of farm information among the farmers varies seasonally. The Table 21 depicts the monthly distribution of farm information broadcasted through selected farm programmes by *All India Radio, Kolkata*.

In case of month wise distribution of farm information broadcasted by radio, table- 30 shows that highest number of farm information was broadcasted in the month of May (112) and month of July (107) came next to it. The highest number of farm information broadcasted in mode of 'Interview' (29) and 'Radio Play' (25) in the month of May. But the highest number of the 'Deliberation' (68) and 'Panel Discussion' type of farm information broadcasted in the month of July as it is found in the study. Average data confirms that farm information under every category was more or less evenly distributed over the months of study period of one year without giving any extra attention to any particular cropping season.

Participation of Experts and Farmers in Radio Programme

Participation of experts often increases the quality of the Farm programme. Farmers can get information directly from the expert and knowledgeable persons. At the same time participation of the farmers and farm people make the farm programme more farmers' oriented and field based problems are addressed properly, farmers can share their experience with their fellow farmers and farm people. As a whole these participation helps the farm programme to become more oriented to bottom line approach. Here, the participation of the experts and others group of people in farm and home programme are computed and presented in the Table 22.

Table 22: Distribution of Presence of Different Group of People in Radio Programme

Group of People Present	*Frequency*	*Per cent*
Experts	156	14.05
Farmers	137	12.34
Farm Women	82	7.39
Rural Youth	56	5.05
General (Anchor Based)	840	75.66
Total Farm Information	1110	100

This Table 22 envisaged that majority of the radio programme (75.66 per cent) broadcasted without presence of any kind of experts or other people except the presenter or anchor. The presence of different group of people along with the anchor was mainly found in case of 'Interview' and 'Panel Discussion' categories of farm information in radio. It is found from the study that experts were present in only 14.05 per cent cases of the farm broadcast, whereas farmers were present in 12.34 per cent cases. Only 7.39 per cent and 5.05 per cent cases of the farm broadcast, farm women and rural youths were present. In number of cases, both experts and farmers or farmers and farm women or farmers and rural youth were present in a single programme.

Allotment of Time in Radio for Broadcasting Farm Information

Like provision of space for publishing any type of farm information in newspapers, allotment of time for the farm information broadcast is also an index through which it can be understood well the extent of importance given to that particular information by the authority. In order to assess importance given by *All India Radio, Kolkata* for broadcasting farm information in terms of allotting time, total time period of farm programme are computed and stratified under different categories, which are depicted in Table 23.

Table 23: Distribution of Time Allotted for Farm Information Broadcast through Radio

Time Allotted	*Frequency*	*Per cent*
Up to 10 minutes	406	36.58
Above 10 minutes to 20 minutes	208	18.74
Above 20 minutes to 30 minutes	452	40.72
Above 30 minutes to 40 minutes	33	2.97
Above 40 minutes to 50 minutes	11	0.99
Total	1110	100

This table shows the distribution of time allotment for farm information broadcast through Radio. It is evident form the study that majority of the farm information in Radio were broadcasted for duration of 20 minutes to 30 minutes (40.72 per cent). 'Up to 10 minutes' (36.58 per cent) came next to it. 18.74 per cent of farm information were also broadcasted in Radio for duration of 10 minutes to 20 minutes.

Farmers' Perception and Preferences of Different Newspapers and Different Radio Programme for Receiving Farm Information

In an effective communication process the responses of the receivers' is one of the most vital component. It helps to understand the need and interest of the receiver and based on which communication process can be made more effective. Here also, in order to farm information through newspapers and radio more effectively, it is necessary to know the need and interest, perception and choice preferences of the farming communities about the farm information. In the following tables the preference of the farmers regarding selection of newspapers out of four selected newspapers and the selection of radio programme out of three radio programmes for getting farm information have been computed and presented.

Table 24: Preference of Newspapers by Farmers for Obtaining Farm Information

Newspapers	*Preference (Per cent)*
Aajkal	6.88
Ananda Bazar Patrika	20.51
Bartaman	48.22
Sambad Pratidin	24.39

The Table 24 ascertains the preference of newspapers for obtaining farm information as responded by the sample farmers taken for the study. *Bartaman* (48.22 per cent) preferred by majority of the farmers as channel for getting farm information among the four newspapers. 24.39 per cent farmers also preferred *Sambad Pratidin* for getting farm information and another 20.51 per cent of farmers preferred

Ananda Bazar Patrika. But only 6.88 per cent farmers preferred *Aajkal* for getting farm information.

In case of getting farm information from radio programme, *Krishi Kathar Asar* ranked the highest (42.63 per cent) in terms of farmers' preference. Followed by *Krishi Kathar Asar*, 38.92 per cent farmers who preferred *Chasi Bhaider Bolchhi* for getting farm information from Radio. Only 18.45 per cent farmers preferred *Uno Jamir Duno Phasal* for getting farm information.

Table 25: Farmers' Preference of Radio Programme for Obtaining Farm Information

Radio Programme	*Preference (Per cent)*
Chasi Bhaider Bolchhi	38.92
Uno Jamir Duno Phasal	18.45
Krishi Kathar Aasar	42.63

Farmers' Perception and Preferences of Different Types Farm Information Communicated by Newspapers and Radio

As it is mentioned in the section 6.3 of this chapter that need and interest of the farm people for farm information should be given proper attention, for that reason farmers' preferences regarding different types of farm information are surveyed, computed and presented in the Tables 26 and 27.

The Table 26 suggests that 32.68 per cent farmers preferred to have farm information in relation to 'Technology Transfer' from selected newspapers. 21.79 per cent farmers preferred to have 'Agricultural Marketing' related farm information followed by 'Rural Development'

(18.56 per cent) and 'Success Story' (15.46 per cent) related farm information in newspapers.

Table 26: Preference of Kind of Farm Information in Newspapers by Farmers

Farm Information	*Preference (Per cent)*
Agricultural Marketing	21.79
Agricultural Policy	8.98
Rural Development	18.56
Success Story	15.46
Technology Transfer	32.68
Weather Forecast	2.53
Total	100

Table 27: Preference of Kind of Farm Information in Radio by Farmers

Farm Information	*Preference (Per cent)*
Agricultural Marketing	7.42
Agricultural Policy	5.27
Rural Development	22.73
Rural Youth	4.31
Success Story	15.63
Technology Transfer	26.48
Weather Forecast	18.16
Total	100

In case of radio Table 27 indicates that 26.48 per cent and 22.73 per cent farmers preferred to have 'Technology Transfer' and 'Rural Development' related farm information respectively from radio programme. Another 18.16 per cent farmers preferred to have weather related information from radio for their farm business followed by 15.63 per cent farmers who preferred to have farm information on different success stories in radio regarding farm business.

Credibility study of Farm Information Communicated through Newspapers and Radio

In the study of farm information, credibility of the farm information is a very important aspect, which needs proper attention. Credibility of farm information delineates its trustworthiness to the farmers and farm people. To study the credibility of farm information communicated by newspapers and radio, both technical (Scientist's/Extension Specialist) and farmers' point of views have been studied.

For the credibility study from technical point of view 200 farm information communicated through four newspapers have been selected randomly in proportion to the number of farm information published in four selected newspapers during the period of study. Same way, 200 farm information communicated through radio has been selected randomly from total number of farm information communicated by *All India Radio, Kolkata* during the period of study. For the credibility study from farmers' point of view, 200 randomly selected farmers of Birbhum district of West Bengal have been selected and interviewed. For credibility study from technical point of view, credibility scores of selected farm information from both newspapers and radio have been computed separately on a 5-point semantic scale ranging from strongly agree to strongly disagree for the fifteen credibility variables (Presentation, Brevity, Detailed, Understandability, Clarity, Necessity, Feasibility, Universality, Timeliness, Current, Repetition, Biased, Accuracy, Factual, Believability). Same way, for credibility study of farm information from farmers' point of view, credibility score have been computed from the farmers' response on credibility of farm information communicated by selected newspapers and radio separately by same credibility scale consisting same fifteen credibility variables. Now the different aspects of credibility study are presented below with the following tables.

Table 28 depicts the mean credibility score of farm information communicated through four newspapers and radio. This has been calculated by taking cumulative credibility score of the same fifteen credibility variables mentioned above. Credibility scores have been computed on same 5-point semantic scale ranging from strongly agree to strongly disagree. Then mean values of credibility scores of farm information for all four selected newspapers and radio have been calculated and presented in the following table:

Table 28: Descriptive Statistics of Credibility Score of Farm Information Communicated through Newspapers and Radio

Mass Media	*Credibility Score*	
	Mean	*Standard Deviation*
Aajkal	54.82	5.87
Ananda Bazar Patrika	54.24	6.94
Bartaman	55.52	6.58
Sambad Pratidin	54.23	6.66
Radio	58.67	6.17

It is evident from the table that *Bartaman* had the highest mean credibility score (55.52) and *Sambad Pratidin* had the lowest (54.23) among the newspapers. But between the two media radio had much higher mean credibility score (58.67) than that of any other newspapers as it is found in the study.

Comparative Study of Credibility of Farm Information communicated by Newspapers from both Technical and Farmers' Point of View

For the present study, credibility of farm information has been studied from both Technical (Scientist/Extension) and Farmers' point of view. In this sub-section comparative study of credibility of farm information communicated by newspapers has been done by the mean values of credibility variables and Factor Analysis. Factor analyses have been done to identify the different factors responsible for

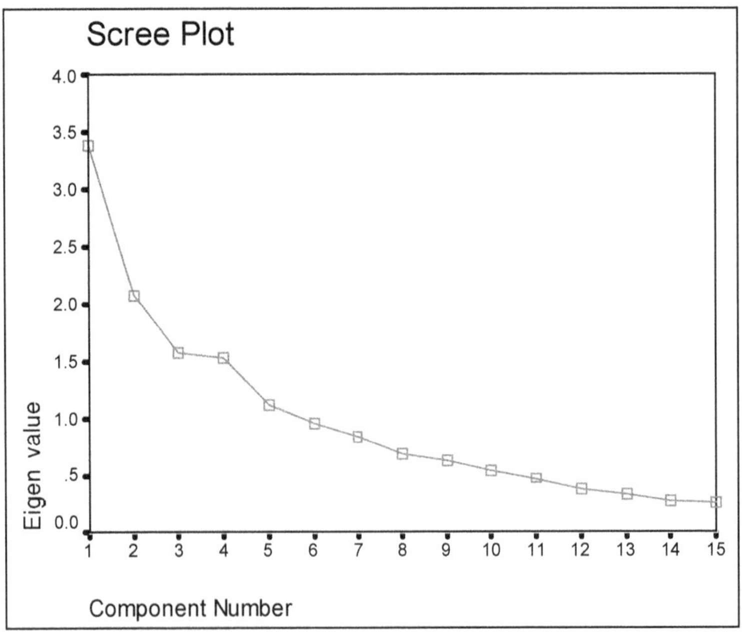

Figure 6: Scree Plot of Credibility Variables Responsible for Credibility of Farm Information Communicated through Newspapers from Technical Point of View

credibility of farm information from both technical and farmers' point of view. At first components were extracted by Principal Component Analysis and those components which have found Eigen Value more than 1 on the Scree Plot were considered major components. Then these components were rotated by Varimax type of rotation of Orthogonal Rotation with Kaiser normalization to obtain Factor Matrix. High Factor Loadings in the rotated factor matrix suggest the different factors responsible for credibility of farm information. Results of these analyses have been presented in the following tables:

Table 29: Descriptive Statistics of Credibility Variables of Farm Information in Newspapers from Technical Point of View

Credibility Variables	*Mean*	*Standard Deviation*	*No. of Farm Information*
Presentation	4.01	0.92	200
Brevity	3.70	0.95	200
Detailed	3.69	1.01	200
Understandability	3.52	1.10	200
Clarity	3.51	1.26	200
Necessity	3.98	0.92	200
Feasibility	4.29	0.82	200
Universality	3.58	1.23	200
Timeliness	3.70	1.03	200
Current	3.73	0.93	200
Repetition	2.81	1.16	200
Biased	3.87	1.07	200
Accuracy	3.77	0.70	200
Factual	3.63	0.92	200
Believability	3.83	0.83	200
Credibility Score	55.58	6.35	200

Table 29 shows the mean values of different credibility variables of farm information communicated through newspapers from technical point of view. It is found in study that 'Feasibility' (4.29 per cent), 'Presentation' (4.01 per cent), 'Necessity' had the higher mean credibility scores on 5- point scale where as Repetition (2.81), Clarity (3.51), Understandability (3.52) had the lower credibility scores in case of newspapers.

Table 30: Total Variance Explained (Newspapers from Technical Point of View)

Component	*Percentage of Variance*	*Cumulative Percentage*
Component 1	17.73	17.73
Component 2	15.12	32.85
Component 3	12.93	45.78
Component 4	10.81	56.60
Component 5	7.84	64.44

To further explore the different underlining dimension responsible for credibility of farm information in newspapers from technical point of view factor analysis was carried out in the study. Scree plot indicates five components were identified in this factor analysis. Table 30 suggests that 64.44 per cent of the variance among the observed variables was determined by five common components.

Table 31 displays the rotated factor matrix of the factor analysis of credibility of farm information published in newspapers from technical point of view. Five-factor solution reveals distinct dimensions. Each factor was composed of some variables, which had high factor loadings and high communalities. The first factor was anchored by

Table 31: Rotated Factor Matrix (Newspapers from Technical Point of View)

Factor	*Variable*	*Factor I*	*Factor II*	*Factor III*	*Factor IV*	*Factor V*	*Commonality*
Factor I	Believability	**0.878**	0.152	0.013	0.083	0.007	0.800
	Factual	**0.804**	0.054	0.057	–0.114	0.037	0.666
	Accuracy	**0.791**	0.134	0.038	0.062	–0.214	0.696
Factor II	Timeliness	0.259	**0.739**	–0.002	0.018	0.008	0.613
	Feasibility	0.173	**0.661**	0.152	0.063	0.278	0.571
Factor III	Presentation	0.082	0.140	**0.874**	–0.045	0.47	0.794
	Detailed	0.109	0.073	**0.855**	–0.082	–0.024	0.756
Factor IV	Understandability	0.100	0.007	0.126	**0.878**	0.003	0.796
	Clarity	–0.047	–0.015	–0.181	**0.876**	–0.028	0.803
Factor V	Current	–0.067	0.031	0.008	–0.054	**0.900**	0.819

'Believability' (0.878) and 'Factual' (0.804). 'Accuracy' (0.791) also loads on the factor. Second factor was made up of 'Timeliness' and 'Feasibility', which had factor loadings 0.739 and 0.661 respectively. The third factor was again built up of two variables, 'Presentation' (0.874) and 'Detailed' (0.855) where as fourth factor was anchored by 'Understandability' (0.878) and 'Clarity' (0.876). Only one variable, Current (0.900) composed the fifth factor.

The Table 32 shows the mean values of different credibility variables of farm information in newspapers from farmer's point of view. According to Table 32 farmers opined that 'Believability' (4.10), 'Current' (4.05) and 'Factual ' (3.99) have the higher mean credibility values where as 'Brevity' (2.25), 'Clarity' (2.69) and 'Feasibility' (2.68) have the lower mean credibility values in case of newspapers.

Table 32: Descriptive Statistics of Credibility Variables of Farm Information in Newspapers from Farmers' Point of View

Credibility Variables	*Mean*	*Standard Deviation*	*No. of Farm Information*
Presentation	3.24	0.99	200
Brevity	2.25	1.15	200
Detailed	3.27	0.79	200
Understandability	3.88	0.63	200
Clarity	2.69	1.04	200
Necessity	3.71	0.57	200
Feasibility	2.68	1.05	200
Universality	3.50	1.00	200
Timeliness	2.83	0.94	200
Current	4.05	0.55	200
Repetition	2.76	0.95	200

Contd...

Table 32–*Contd...*

Credibility Variables	*Mean*	*Standard Deviation*	*No. of Farm Information*
Biased	3.17	0.91	200
Accuracy	3.66	0.61	200
Factual	3.99	0.48	200
Believability	4.10	0.60	200
Credibility Score	49.72	4.23	200

To identify different underlining dimensions associated with the credibility of farm information in newspapers from farmer's point of view, factor analysis has been done for the purpose with same fifteen credibility variables. Figure 7 suggests six components had Eigen value more than 1 and hence, those components were extracted with Principal Component Analysis. The Table 33 ascertains that those six factors accounted for 75.29 per cent of variance.

Table 33: Total Variance Explained (Newspapers from Farmers' Point of View)

Component	*Percentage of Variance*	*Cumulative Percentage*
Component 1	17.65	17.65
Component 2	13.14	30.79
Component 3	12.74	43.53
Component 4	11.67	55.20
Component 5	11.18	66.38
Component 6	8.91	75.29

Table 34 shows the rotated factor matrix of the factor analysis of credibility of the farm information in newspapers from farmers' point of view. Six factors were found have distinct dimensions. First factor was made up of 'Repetition'

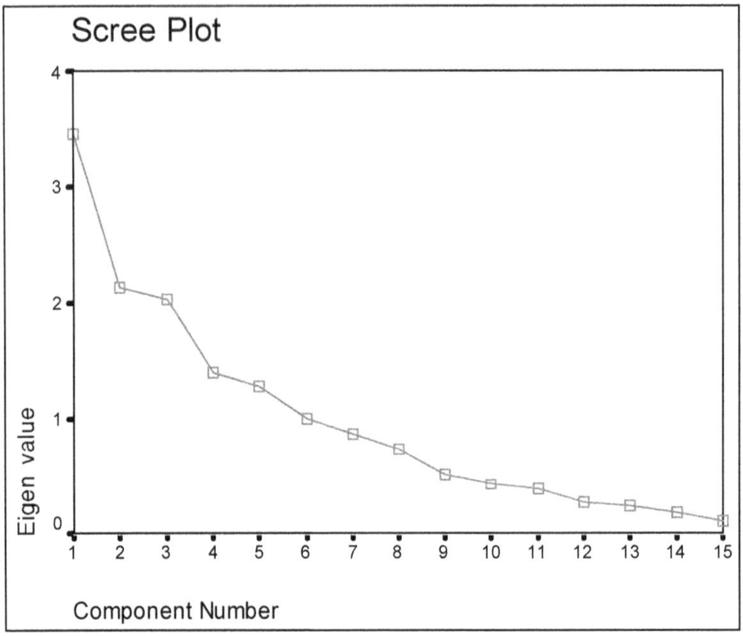

Figure 7: Scree Plot of Credibility Variables Responsible for Credibility of Farm Information Communicated through Newspapers from Farmers' Point of View

(0.774), 'Necessity' (0.774) and 'Detailed' (0.665). Second factor was anchored by 'Universality' (0.898) and 'Brevity' (0.746). 'Believability' (0.645) also had significant loadings on the second factor. Third and sixth factors were built up with only one variable each. They were made up of 'Feasibility' (0.823) and 'Factual' (0.928) respectively. Fourth factor centered on 'Clarity' (0.849) and 'Presentation' (0.592) whereas fifth factor centered on 'Accuracy' (0.825) and 'Biased' (0.726).

So, in this comparative study some differences has been found between technical and farmers' point of view in terms of credibility of farm information communicated by four

Table 34: Rotated Factor Matrix (Newspapers from Farmers' Point of View)

Factor	*Variable*	*Factor I*	*Factor II*	*Factor III*	*Factor IV*	*Factor V*	*Factor VI*	*Commonality*
Factor I	Repetition	**0.774**	0.061	–0.194	0.169	0.347	–0.030	0.791
	Necessity	**0.774**	0.007	0.072	0.022	–0.131	0.039	0.624
	Detailed	**0.665**	0.182	0.418	0.118	–0.182	–0.087	0.705
Factor II	Universality	–0.113	**0.898**	0.036	–0.194	–0.019	–0.183	0.892
	Brevity	0.129	**0.746**	–0.188	0.406	0.002	0.153	0.797
	Believability	0.280	**0.645**	0.045	0.014	0.398	0.155	0.680
Factor III	Feasibility	–0.168	–0.076	**0.823**	–0.122	0.040	0.143	0.748
Factor IV	Clarity	–0.037	–0.080	–0.007	**0.849**	–0.068	0.177	0.764
	Presentation	0.520	0.244	–0.167	**0.592**	0.284	–0.208	0.832
Factor V	Accuracy	–0.024	–0.013	0.226	–0.247	**0.825**	–0.098	0.802
	Biased	–0.195	0.221	–0.275	0.338	**0.726**	0.060	0.808
Factor VI	Factual	–0.031	–0.023	–0.009	0.103	–0.003	**0.928**	0.873

selected newspapers which have been discussed elaborately in the discussion part of this Chapter.

Comparative Study of Credibility of Farm Information Broadcasted through Radio from both Technical and Farmers' Point of View

Like newspapers, comparative study of credibility of farm information has also been studied from both Technical (Scientist/Extension) and Farmers' point of view. The same methods as mentioned earlier have also been applied here. Results of these analyses have been presented in the following tables:

Table 35: Descriptive Statistics of Credibility Variables of Farm Information Broadcasted through Radio from Technical Point of View

Credibility Variables	*Mean*	*Standard Deviation*	*No. of Farm Information*
Presentation	4.27	0.73	200
Brevity	3.84	0.90	200
Detailed	3.96	0.83	200
Understandability	4.22	0.90	200
Clarity	3.76	0.94	200
Necessity	3.28	1.42	200
Feasibility	4.20	0.82	200
Universality	3.65	1.41	200
Timeliness	3.57	1.18	200
Current	3.25	0.92	200
Repetition	3.87	1.15	200
Biased	4.09	1.10	200
Accuracy	4.04	0.80	200
Factual	4.23	0.62	200
Believability	4.25	0.67	200
Credibility Score	58.44	6.45	200

The Table 35 displays the mean credibility values of fifteen credibility variables of farm information communicated through radio from technical point of view. It was found in the study that 'Presentation' (4.27), 'Believability' (4.25), 'Factual' (4.23) had the higher mean credibility scores where as 'Current' (3.25), 'Necessity' (3.28), 'Universality' (3.65) had the lower mean credibility values.

Table 36: Total Variance Explained (Radio from Technical Point of View)

Component	*Percentage of Variance*	*Cumulative Percentage*
Component 1	20.81	20.81
Component 2	17.78	38.59
Component 3	11.30	49.89
Component 4	10.59	60.48
Component 5	8.23	68.72

Like newspapers, factor analysis of credibility of farm information in radio from technical point of view has been done to explore the underlining dimensions. Scree plot suggested that five components were identified having Eigen value more than 1. Hence five factors had distinct dimensions in this case, which have been extracted through Principal Component Analysis. Table-36 ascertains that these five factors were accounted for 68.72 per cent variance.

Table 37 displays the rotated factor matrix of the factor analysis of the credibility of farm information in radio from technical point of view. It ascertains that five factors had distinct dimensions. First factor here also anchored by 'Believability' (0.815) and 'Accuracy' (0.815) like

Table 37: Rotated Factor Matrix (Radio from Technical Point of View)

Factor	*Variable*	*Factor I*	*Factor II*	*Factor III*	*Factor IV*	*Factor V*	*Commonality*
Factor I	Believability	**0.815**	0.054	0.031	0.113	–0.188	0.716
	Accuracy	**0.815**	–0.086	–0.035	–0.348	0.001	0.794
	Biased	**0.759**	0.003	0.064	0.052	–0.191	0.620
Factor II	Timeliness	0.169	**0.882**	–0.036	–0.001	0.111	0.820
	Necessity	0.261	**0.803**	–0.142	–0.081	0.249	0.802
	Feasibility	–0.056	**0.769**	0.196	0.136	–0.213	0.697
Factor III	Detailed	–0.002	–0.091	**0.908**	0.032	0.078	0.840
	Presentation	0.072	0.144	**0.854**	–0.082	–0.082	0.769
Factor IV	Understandability	0.188	0.268	–0.020	**0.857**	0.137	0.861
	Clarity	–0.049	–0.214	–0.040	**0.813**	–0.279	0.789
Factor V	Current	–0.106	0.079	0.032	–0.080	**0.764**	0.609

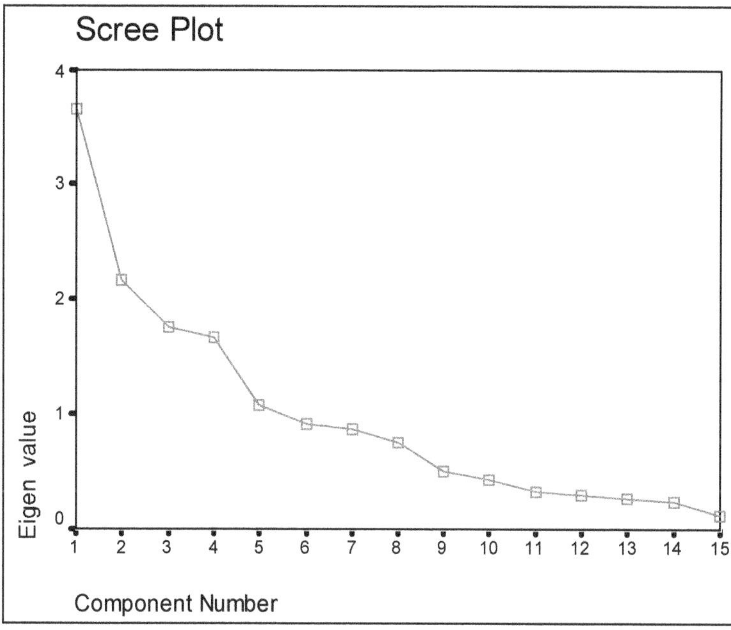

Figure 8: Scree Plot of Credibility Variables Responsible for Credibility of Farm Information Communicated through Radio from Technical Point of View

newspapers. 'Biased' (0.759) had also load in the factor. Second factor was composed of 'Timeliness' (0.882), 'Necessity' (0.803) and 'Feasibility' (0.769). Third factor was made up of two variables, 'Detailed' and 'Presentation'. They have the factor loads of 0.908 and 0.854 respectively. Fourth factor was anchored by 'Understandability' (0.857) and 'Clarity' (0.813). The fifth factor was composed of only one variable, which was 'Current' and it had load of 0.764.

The Table 38 ascertains the mean values of different credibility variables of farm information communicated through radio as perceived by farmers. It is found that 'Understandability' (4.43), 'Repetition' (4.20), 'Believability'

(4.19). 'Factual' (4.18) had the higher mean values, whereas 'Feasibility' (3.20), 'Timeliness' (3.20), 'Detailed' (3.78) etc. had the lower mean values.

Table 38: Descriptive Statistics of Credibility Variables of Farm Information in Radio from Farmers' Point of View

Credibility Variables	*Mean*	*Standard Deviation*	*No. of Farm Information*
Presentation	3.94	0.77	200
Brevity	3.40	1.02	200
Detailed	3.76	0.77	200
Understandability	4.43	0.60	200
Clarity	3.78	0.90	200
Necessity	4.00	0.65	200
Feasibility	3.20	0.98	200
Universality	4.50	0.50	200
Timeliness	3.20	1.05	200
Current	3.84	0.43	200
Repetition	4.20	0.59	200
Biased	4.16	0.76	200
Accuracy	4.03	0.71	200
Factual	4.18	0.62	200
Believability	4.19	0.43	200
Credibility Score	58.77	4.69	200

To explore the underlining dimensions of credibility of farm information in radio according to farmers' opinion, factor analysis has also been done here. Figure 9 displays that five components had Eigen value more than one. Through Principal Component Analysis it was found that these five factors were accounted for 76.13 per cent of variance as it can be seen in Table 39.

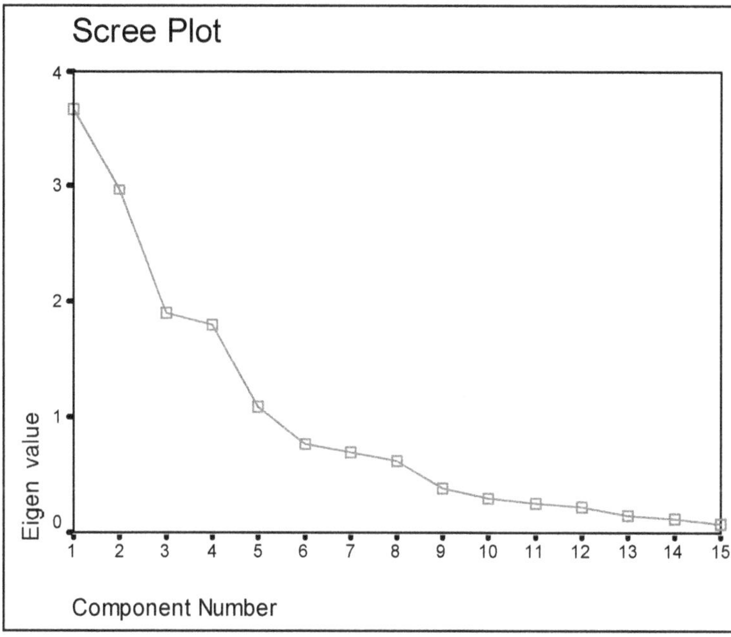

Figure 9: Scree Plot of Credibility Variables Responsible for Credibility of Farm Information Communicated through Radio from Farmers' Point of View

Table 39: Total Variance Explained (Radio from Farmers' Point of View)

Component	*Percentage of Variance*	*Cumulative Percentage*
Component 1	21.36	21.36
Component 2	16.73	38.10
Component 3	15.84	53.94
Component 4	12.92	66.85
Component 5	9.28	76.13

This Table 40 shows the rotated factor matrix of the credibility of farm information in radio as perceived by the farmers. Here the first factor was anchored by 'Timeliness'

Table 40: Rotated Factor Matrix (Radio from Farmers' Point of View)

Factor	*Variable*	*Factor I*	*Factor II*	*Factor III*	*Factor IV*	*Factor V*	*Commonality*
Factor I	Timeliness	**0.811**	0.239	–0.039	0.042	0.202	0.758
	Biased	**0.802**	–0.208	–0.070	0.169	–0.158	0.745
	Universality	**0.729**	0.108	0.152	–0.124	0.282	0.661
	Factual	**0.663**	0.081	–0.665	–0.013	–0.019	0.890
Factor II	Repetition	–0.157	**0.714**	0.087	0.558	–0.112	0.867
	Presentation	0.303	**0.686**	0.192	0.135	–0.239	0.674
	Detailed	0.477	**0.651**	–0.132	–0.246	–0.188	0.764
Factor III	Brevity	0.041	–0.004	**0.897**	0.039	–0.201	0.848
	Clarity	–0.134	0.278	**0.750**	0.273	0.052	0.734
	Believability	0.526	0.087	**0.647**	–0.048	0.287	0.787
Factor IV	Necessity	0.150	0.016	0.059	**0.833**	0.121	0.734
Factor V	Accuracy	0.173	–0.100	–0.047	–0.073	**0.913**	0.882

(0.811) and 'Biased' (0.802). 'Universality' (0.729) and 'Factual' (0.663) had also load on the first factor. Second factor was composed of 'Repetition' (0.714), 'Presentation' (0.686) and 'Detailed' (0.651) and third factor was made up of 'Brevity' (0.897), 'Clarity' (0.750), and 'Believability' (0.647). Forth and fifth factors were made up of only one variable each. They were 'Necessity' (0.833) and 'Accuracy' (0.913) respectively.

Relationship between Credibility Scores and Social Variables

Here, an attempt was made to find out relationship among credibility score and social variables from the farmers' responses for both newspapers and Radio. For the purposes linear correction co-efficient was computed and presented in the Table 41.

Table 41: Correlation Co-efficient of Credibility Scores and Social Variables

Variables	*Credibility Score (Newspapers)*	*Credibility Score (Radio)*
Educational Status	0.702**	0.432*
Family Income	0.112	0.131
Family Type	0.221	0.179
Family Size	0.071	0.106
Holding Status	0.476*	0.598**
Social Participation	0.534**	0.672**
Exposure to Channels	0.687**	0.726**

*: 5 per cent level of Significance; **: 1 per cent level of Significance

Table 41 suggests that educational status, social participation and Exposure to Channels of Farm

Information Communication had positive and significant relationship both at 5 per cent, 1 per cent level of significance with credibility scores of farm information both in newspapers and radio. In case of farm information published in newspaper, holding status had positive and significant relationship with credibility score at 5 per cent level of significance but in case of radio it was significant both at 5 per cent and 1 per cent level of significance.

Difference between Cumulative Credibility Score of Farm Information Communicated through Newspapers and Radio

To find out existence of any significant difference between cumulative credibility scores of farm information communicated through newspaper and radio from technical point of view paired *t*-test was employed. Results of this test have been presented in Table 42.

In a 2 tailed test it is found that mean credibility value were not equal both at 5 per cent and 1 per cent level of significance. Estimated mean difference was 2.860 and mean value was higher in case of radio. Same results have been found when paired *t*-test was carried out between cumulative credibility scores of farm information communicated through newspapers and radio from framers' point of view. Here also mean values were significantly (both at 5 per cent and 1 per cent level of significance) not equal as the Table 42 suggests. The estimated mean difference was here 9.050 and mean credibility value was higher in case of radio, rather than newspapers.

Table 42: Paired *t*-Test between Cumulative Credibility Score of Farm Information Communicated through Newspapers and Radio

Point of Views	*Mean Value of Farm Information (Radio)*	*Mean Value of Farm Information (Newspapers)*	*Estimated Mean Difference*	*No. of Observation*	*t-Test Value*	*Table Value (1% level, 199 d.f)*
Technical	58.435	55.575	2.860	200	4.41	2.576
Farmers	58.770	49.720	9.050	200	19.87	2.576

Discussion

The result of the study has helped to enlighten over the different research questions taken for the study. Different results show us the overall picture of several aspects of farm information dissemination by mass media, like newspaper and radio in West Bengal, India as well as farmers' perception regarding those mass media. In this section, discussion of the obtained results has been presented objective wise:

Objective-I

- ✰ To study the comparative orientation and importance given to farm information by the selected newspapers and radio programme.

As it is evident from the Table 43 that regarding different items of assessment like, number of publication during the study period; space covered; title font size; preferential selection of major categories and sub-categories; monthly distribution, readability- *Bartaman* was found best in most of the cases. It speaks in favour of the fact that newspapers had differential orientation towards farm information as well as, they give differential importance towards farm information, which satisfy the objective-I.

Same results were found by Roy *et al.* (2003). They found in their study that the coverage of farm news was somewhat satisfactory in Ganashakti, Bartaman and Sambad Pratidin while it was altogether unsatisfactory in *Ananda Bazar Patrika* and *Aajkal*. Marketing aspects, natural hazards, agricultural polity and social life received more importance in those newspapers.

Table 43: Results of Comparative Study of Selected Newspapers

Item Assessed	*Newspapers*				*Table No.*	*Interpretation*
	Aajkal	*Ananda Bazar Patrika*	*Bartaman*	*Sambad Pratidin*		
No. of publication during study year	Rank-IV	Rank-II	Rank-I	Rank-III	5	*Bartaman* gave comparatively more importance towards farm information communication
Space provided for farm information to total space published	Rank-IV	Rank-II	Rank-I	Rank-II	6	In comparison to total space published, provision for farm information was very poor, which manifest lesser importance given by all newspapers. However *Bartaman* was again best one
Distribution of space provided	Up to 50 sq.cm. and Above 50-100 sq.cm.	Above 100-150 sq.cm. and Above 50-100 sq.cm.	Above 50-100 sq.cm. and Up to 50 sq.cm.	Above 50-100 sq.cm and Above 100-150 sq.cm.	13	All the newspapers are more or less homogenous in nature
Distribution of font size of seadlines	31-40 and 41-50	41-50 and 31-40	31-40 and 41-50	41-50 and 31-40	12	All the newspapers are more or less homogenous in nature and major font size of Headlines were 31-40 and 41-50
Month wise distribution of farm information	No major variation was found				11	No as such seasonality variation was found but Bartaman give little importance to *Boro* season

Contd...

Table 43–*Contd...*

Item Assessed	*Newspapers*				*Table No.*	*Interpretation*
	Aajkal	*Ananda Bazar Patrika*	*Bartaman*	*Sambad Pratidin*		
Major farm publication in descending order	News, Views and Advertisement with lion share of News	News, Views and Advertisement with lion share of News	News, Views and Advertisement with balanced publication	News, Views and Advertisement with lion share of News	7	News is the major category of farm information but *Bartaman* maintained more balance
Monthly distribution of major categories	News were preferred for all months by all newspapers				11	All four newspapers were homogenous
Preferential selection of major three sub-categories in descending order	News (Agril. Pollution, Agril. Marketing, Crisis); Views (Crisis, Agril. Pollution, Agril. Marketing); Advertise. (Co-operative)	News (Agril. Policy, Crisis, Rural Dev.); Views (Agril. Policy, crisis, Tech. Transfer, Rural Dev.); Advertise. (Agril. Machinery, Fertilizer, Rural Dev.)	News (Agril. Policy, Crisis, Weather Forecast); Views (Tech. Transfer, Crisis, Agril. Policy); Advertise. (Fertilizer, Agril. Machinery)	News (Agril. Policy, Crisis, Agril. Marketing); Views (Tech. Transfer, Crisis, Agril. Policy); Advertise. (Plant Protection Material, Ruarl Dev.)	8,9,10	No major difference in case of News and Views and in case of Advertisements common tilt towards Agril. Machinery and Fertilizer.
Readability	Above Standard	Above Standard	Standard	Standard	14	*Bartaman* and *Sambad Pratidin* were standard for Farm People and rest two were of higher stanadrd

Table 44: Results of Comparative Study of Selected Radio Programme

Item Assessed	*Categories of Farm Information Broadcasted through Selected Radio Programme*				*Table No.*	*Interpretation*
	Deliberation	*Interview*	*Panel Discussion*	*Radio Play*		
No. of farm information	I	II	IV	III	16	Deliberation is the mode of presentation preferred by all three radio progarmme and no variation observed
Time allotted	Time allotted of 20 to 30 minutes for farm information broadcast was found most preferred by radio followed by up to 10 minutes slot programme				23	Preference for 20 to 30 minutes was found predominant
Monthly variation	No variation but preference to July	No variation but preference to May	No variation but preference to July	No variation but preference to May	21	No major variation was found
Preferential selection of major sub-categories	Technology Transfer, Rural Youth, Farm Women were the major sub-categories for all four categories of farm information broadcasted through radio				17, 18, 19,20	Technology Transfer, Rural Youth, Farm Women were preferred for farm broadcast
Participation of different group of people	Rank-I. Rank-II Rank-III	- - -	Anchor based Expert are involved Farmers are Involved		22	All the progarmme under study had preferences for involving different group of people in their programme

In case of radio, another, another selected mass media for the present study, no major different were found regarding different item of assessment for all the three studied programme, like; *Chasi Bhaider Bolchhi, Uno Jamir Duno Phasal* and *Krishi Kathar Aasar*. In case of time allotment for different farm programme, it was observed that majority of the farm information broadcasted in Above 20 minutes to 30 minutes and up to 10 minutes time slots. This is quite reasonable because farm people often do not get enough time in the morning and afternoon hours due to field and household activities. Longer time slot programmes are justifiably broadcasted during evening hours when farm people get enough time after their work, which has been observed during the study.

Newspapers and radio being the most popular mass media, as it is observed from the results, communicate farm information regularly. But comparatively radio was found better communicator of farm information than that of all the four selected newspapers. It can also be observed that these media had their own preference for selection of different major categories and sub-categories of farm information. Moreover no rule or method could be traced; how and why these two media have preferred these categories and sub-categories. Even there was no evidence found that the print media under study considered the requirements of farmers' need for farm information. In this respect, radio programmes were found much better than that of newspapers.

So far as the shouldering social responsibility of communicating of farm information is concerned radio, were found much better than that of newspapers.

Objective-II

☆ To study the farm information seeking behaviour of the sample farmers.

So far as farmers' the preference of newspapers for obtaining farm information is concerned, *Bartaman* was ranked first followed by *Sambad Pratidin* and *Ananda Bazar Patrika* (Table 24). In the earlier section also *Bartaman* was found best newspapers communicating farm information.

In case of preferences for kind of farm information by the farmers published in selected newspapers, it was observed (Tables 26 and 27) that farmers obtained 'Technology Transfer', 'Agricultural Marketing', and 'Rural Development', 'Success Story' related farm information from newspapers. But content analysis suggests that all the newspapers publish 'Agricultural Policy', 'Crisis' related farm information more. Whereas in case of radio, farmers again obtained 'Technology Transfer', 'Rural Development', 'Success Story' etc related farm information more. Content analysis of farm information in radio shows that radio broadcasted farm information in this particular area more. So it indicates that *All India Radio, Kolkata* has better understanding of need and interest of the farmers and farm people and their behavioural pattern in terms of obtaining farm information.

In case of radio (Table 25), *Krishi Kathar Aasar* was the most preferred radio programme by the farmers followed by *Chasi Bhaider Bolchhi. Uno Jamir Duno Phasal* was least preferred radio programme for obtaining farm information. In case of radio programme, majority of the farmers gathers farm information from *Krishi Kathar Asar*. This is mainly because; this programme is broadcasted in the evening hours when farmers get enough time, which is mentioned earlier. A large percentage of farmers also gather information in morning hours before going out for their work from *Chasi Bhaider Bolchhi*. But a very less percentage of farm people can attend the programme *Uno Jamir Duno Phasal*, broadcasted in Afternoon. Because most of them are busy at that time in their work.

Objective-III and IV

- ☆ To study the credibility of farm information communicated through newspapers and radio from technical point of view.
- ☆ To study the credibility of farm information communicated through Bengali newspapers and radio as perceived by the farmers.

In the present study an attempt has been made to study the credibility of farm information communicated through newspapers and radio. Credibility study is a complex and multidimensional concept. Different researchers employed different techniques to study the media credibility earlier. After an exhaustive study of existing literature, credibility of farm information communicated by these two media has been quantified in 5-point semantic scale. Credibility of farm information has been compared both in newspaper and radio from technical point of view and farmers' point of view. To handle the multidimensional nature of media credibility, factor analysis was carried out. The mean credibility scores again tell the same story that farm information published in *Bartaman* had the highest mean credibility score than that of any newspapers. Between the two media, farm information communicated through radio had much higher mean credibility than that of newspapers.

This is done from technical point of view on 200 sample farm information drawn both from communicated by newspapers and radio which is discussed earlier.

Now the factor matrix of the credibility of farm information published in newspaper from technical point of view depicts that the most important first factor was anchored by 'Believability', 'Factual' and 'Accuracy'. But from farmers' point of view first factor is made up of 'Repetition', 'Necessity' and 'Detailed'. Hence, it is evident that farmers have different perception regarding credibility of farm information in newspapers. They stressed on repetition of farm information, its necessity to them and how detailed the information were published. But 'Believability', 'Factual' and 'Accuracy' were not the foremost criteria to them for credibility of farm information in newspapers. The anomaly was again found in credibility of farm information broadcasted through radio from technical and farmers' point of view. Here, once again, first factor was constructed by 'Believability' and 'Accuracy'. 'Biased' also had load on the first factor. But from farmers' point of view, the first credibility factor of farm information communicated through radio was built up of 'Timeliness', 'Biased', 'Universality' and 'Factual'. Once again farmers were found not too much concerned about 'Believability' and 'Accuracy' of farm information regarding their perception on credibility of farm information broadcasted through radio.

Objectives-V and VI

- ✰ To enumerate the relationship of different socio-economic and socio-personal variables with the credibility of farm information.

- ☆ To study the differential perception of credibility of farm information communicated through selected newspapers and radio programme from technical and farmers' point of view and:

It is found in the study that educational status, social participation, holding status and Exposure to Channels of Farm Information Communication had positive and significant relationship with the credibility score of farm information communicated through newspapers and radio. Hence, it can be stated that a positive change in these variables enhances the credibility of farm information as perceived by the farmers.

In Paired *t*-tests, in both, technical and farmers' point of view, it was found that farm information communicated through radio had higher mean cumulative score than that of newspapers. Hence, it can be generalized that farm information communicated through radio has appeared to be more credible than that of newspapers, from both technical and farmers point of view in terms of dissemination of farm information as perceived by the farmers.

As a whole, it can be generalized from the results of the study that *All India Radio, Kolkata* was found more concerned about dissemination of farm information between the two media; whereas, among the newspapers, *Bartaman* shouldered relatively better role in terms of dissemination of farm information to the farm people. Moreover, the results depict that need for the farm information of the farmers were not taken care of by the newspapers whereas, this gap was relatively less in case of radio programmes. In terms of credibility of farm information, again the difference was quite prominent. But between the two media, radio was found more credible in terms of dissemination of farm information.

Chapter 5

Learning Experience

In India, with the ever increasing population, our food demand is also increasing day by day. As the scope of horizontal growth of development is very limited, we have to rely on vertical development of agriculture. For vertical development of agriculture, farm information communication has a pivotal role to play for the speedy communication and subsequent adoption of our research findings in the field of agriculture by our farmers. In case of farm information communication, the potentiality of the mass media is still not exploited effectively in India. Very recently, Government of India has taken initiative to provide mass media support to agricultural extension. Among the different mass media, newspapers and radio have some relative advantages in terms of cost-effectiveness and wide coverage. Television as a media and a channel for the dissemination has tremendous potentiality and efficacy. But Television has not been taken for the present research because its lack of relevance in the study area. In West Bengal, several newspaper dailies are being published every day. Among those, most widely circulated are *Ananda Bazar Patrika*, *Aajkal*, *Bartaman* and *Sambad Pratidin*. These newspapers publish farm information in Bengali Language. In the same way *All India Radio, Kolkata* broadcasts farm information in three different time slots in Bengali everyday. But different aspects of these farm information published by those four newspapers and broadcasted by *All India Radio, Kolkata* needs to be scrutinized. At the same time, opinion of the farm people regarding farm information, disseminated by these two media and needs and interest of the farm people should be enlightened. Hence, in this backdrop after exhaustive study of existing literature present study has been formulated in an exploratory research design with the following objectives:

- ✰ To study the comparative orientation and importance given to farm information by the selected newspapers and radio programme.
- ✰ To study the farm information seeking behaviour of the sample farmers.
- ✰ To study the credibility of farm information communicated through newspapers and radio from technical point of view.
- ✰ To study the credibility of farm information communicated through Bengali newspapers and radio as perceived by the farmers.
- ✰ To enumerate the relationship of different socio-economic and socio-personal variables with the credibility of farm information.
- ✰ To study the differential perception of credibility of farm information communicated through selected newspapers and radio programmes from technical and farmers' point of view and;
- ✰ To suggest some policy measures.

In the present study, analyses of the content of the farm information communicated by four newspapers have been done. The farm information communicated through newspapers have been categorized into three different category; News, Views and Advertisements. Every category is again subdivided in to different sub-categories according to the type of farm information published in different newspapers. Different categories and sub-categories have been tested through reliability test. Krippendorff's alpha was found 0.832, which is permissible, and subject under study was found reliable. Frequency and percentage distribution of these categories and sub-categories have been studied. Space covered by the farm information was measured in sq. cm. and space covered by individual farm information was categorized in ten different strata for measuring frequency distribution. Font size of the headlines of the farm information was measured with the help of existing Bengali Software Packages and they were also categorized in ten different strata for measuring frequency distribution.

In case of radio, three farm programmes; *Chasi Bhaider Bolchhi* (Morning), *Uno Jamir Duno Phasal* (Afternoon) and *Krishi Kathar Aasar* (Evening) broadcasted through *All India Radio, Kolkata* at 456.6m. Band and 657 KHZ were studied and recorded. One year study period was taken. Like newspapers, for the analysis of content, farm information was divided under four broad categories of 'Deliberation', 'Interview', 'Radio Play' and 'Panel Discussion'. Every category was again subdivided into different sub-categories. Reliability test with Krippendorff's alpha was computed and value of alpha was found 0.817, which is permissible. Participation of other people in broadcasting farm programmes other than presenter or anchor was also studied for Radio. Time allotted for Radio for each farm information was measured in minutes and were categorized into five different strata for measuring frequency distribution. In both newspapers and radio month wise distribution of farm information were also studied. For credibility study of farm information communicated through newspapers and radio from technical point of view, two hundred sample farm information have been selected from four newspapers through Random Sampling Method (without replacement) in the proportion of farm information published in those newspapers. In the same way, two hundred sample farm information were selected from the total farm information broadcasted in radio through Random Sampling Method (without replacement). For credibility study of farm information communicated through newspapers and radio from technical point of view, fifteen credibility parameters were selected and were measured with the help of a modified five point semantic scale (ranging from strongly agree to strongly disagree) developed by Gaziano and McGrath (1986). Reliability of the scale was measured for the farm information communicated through newspapers and radio by Internal Consistency Method by measuring Cronbach's Alpha (Cronbach, 1951) and it was found 0.746, which was accepted as reliable. Now for the sample farmers' orientation two hundred farmers of Birbhum district of West Bengal were interviewed with structured schedule. The state and district were selected purposively. Respondents were selected through cluster sampling method from nineteen blocks of the district in proportion to the number of cultivator present in every block. Different socio-economic variables of the sample farmers like–Educational Status, Family Size, Family Type, Holding Size, and Social Participation were measured by the 'Socio-Economic Status Scale-Rural' developed by Pareek and Trivedi (1964). Exposure to Channels of Farm Information Communication was measured by the help of the scale developed by Bharadwaj (1988). Different preferences of the farmers for obtaining farm information were also studied. For credibility study of farm information from farmers' point of view, the same credibility scale was applied. Reliability score measured for the credibility scale on fifty farmers applying Internal Consistency Method and it was measured 0.782, which is permissible. Data have been analyzed with Frequency and Percentage Distribution, Mean Distribution, Standard Deviation, Co-efficient of Correlation. To analyze the different underlining dimensions of credibility of form information Factor Analysis was employed. Paired *t*-test was also employed to measure mean differences.

After an intensive study, it is clear that performance of Bartaman was the best among all the newspapers, but between the two media, radio was found far better. But total percentage coverage by farm information in newspapers in relation to other information was very poor. This is also true in case of radio. *Bartaman* ranked the highest in this regard, but still it is 1.01 per cent. In case of radio it was only 7.09 per cent. It is evident from the study that no as such, seasonality was found among the four selected newspapers in terms of dissemination of farm information. Farm information were more or less evenly distributed in every months, of the study period. Only *Bartaman* gave little bit extra attention to the *Boro* season.

All the newspapers published most of the farm information under 'News' category. But *Bartaman* maintained more balance in terms of 'News', 'Views' and 'Advertisements'. All the four newspapers emphasized 'Agricultural Policy', 'Crisis', 'Agricultural Marketing' and 'Rural Development' type of farm information under 'News' category. But in case of 'Views' type farm information, *Bartaman* and *Sambad Pratidin* gave much more importance to the 'Technology Transfer' sub-category. At the same time 'Agricultural Policy', 'Crisis', 'Agricultural Marketing' were also given importance in all the four newspapers under views category. In case of 'Advertisements'; *Aajkal* emphasized on 'Cooperative', *Ananda Bazar Patrika* emphasized on 'Agricultural Machinery', *Bartaman* emphasized on 'Fertilizer' and 'Agricultural Machinery' and *Sambad Pratidin* emphasized on 'Plant Protection Materials'. In all the newspapers, in most of the cases headlines of the farm information published using font size ranging from 31 to 50. In *Aajkal, Ananda Bazar Patrika* and *Sambad Pratidin,* space provided for most of the farm information are up to 150 sq. cm. Only in *Bartaman,* most of the farm information spaced up to 200 sq. cm. *Sambad Pratidin* and *Bartaman* were found standard newspapers to the farmers in terms of readability of the farm information.

In case of *Radio,* most of the farm information was broadcasted under the broad category of 'Deliberation'. Here also, no as such seasonality was observed in terms of dissemination farm information under different categories of farm information broadcasted through radio. Farm information were more or less evenly distributed in all the months of the study period.

The study envisaged that majority of the radio programme (75.66 per cent) broadcasted without presence of any kind of experts or other people except the presenter or anchor. The presence of different group of people along with the presenter is mainly found in case of Interview and Panel Discussion categories of farm information in radio.

In terms of disseminating farm information under different categories, 'Deliberation' were mainly concentrated on 'Technology Transfer' and 'Rural Development' sub-categories; 'Interview' on 'Technology Transfer', 'Health and Sanitation' and 'Rural Development' sub-categories; 'Panel Discussion' on 'Rural Youth' and 'Farm Women' sub-categories and 'Radio Play' concentrated on 'Rural Development', 'Success Story' and 'Technology Transfer' sub-categories.

Most of the farm information has been broadcasted in Radio in 20 minutes to 30 minutes time slot. A sizable numbers of farm information were also broadcasted up to 10 minutes time slot.

Now the farmers' responses have been taken in this study regarding dissemination of farm information in newspapers and radio. Farmers preferred most the *'Bartaman'* as a newspaper for obtaining farm information and in case of Radio they preferred *Krishi Kathar Asar* and *Chasi Bhaider Bolchhi*. In case of type of farm information, farmers mostly sought farm information regarding 'Transfer of Technology', 'Rural Development', 'Agricultural Marketing' and 'Weather Forecast' from newspapers and radio.

A credibility study of 200 randomly selected (by proportional allocation) farm information from newspapers and also 200 farm information randomly selected from farm information broadcasted by *All India Radio, Kolkata* during study was conducted from technical point of view. At the same time, credibility study of farm information communicated by newspapers and radio from farmers' point of view was also recorded. In terms of credibility of farm information from technical point of view *Bartaman* had the highest mean credibility score and *Radio* ranked highest between the media. From technical point of view, in newspapers; 'Feasibility', 'Presentation' and 'Necessity' had the highest mean credibility score. But from farmers' point of view in newspapers; 'Believability', 'Current' and 'Factual' had the highest mean credibility score. Now the factor matrix of the credibility of farm information published in newspaper from technical point of view depicts that the most important first factor was anchored by 'Believability', 'Factual' and 'Accuracy'. But from farmers point of view, first factor was made up of 'Repetition', 'Necessity' and 'Detailed'. Now in case of Radio from technical point of view; 'Presentation', 'Believability' and 'Factual' had the highest mean credibility score and from farmers' point of view in Radio; 'Understandability', 'Repetition', 'Believability' and 'Factual' had the highest mean credibility score. From technical point of view in Radio, once again first factor was constructed by 'Believability' and 'Accuracy'. 'Biased' had also load on the first factor. But from farmers' point of view the first credibility factor of farm information communicated through radio was built up of 'Timeliness', 'Biased', 'Universality' and 'Factual'. It was found in the study that educational status, social participation, holding status and exposure to channels of farm information communication had positive and significant relationship with the credibility score of farm information communicated through newspapers and radio. Cumulative credibility scores were found significantly higher in case of Radio than that of newspapers from both technical and farmers' point of view.

So, from the study, it can be concluded that the status of dissemination of farm information both in newspapers and radio is not so good. These two important media have lot of place to improve their condition in terms of farm information communication. First of all, number of information is needed to be increased in these two media and newspapers in particular. Among the newspapers, except *Bartaman,* all other newspapers are not shouldering the responsibility in terms of disseminating information to the farm people. Proper balance is needed in terms of 'News', 'Views' and 'Advertisements' in case of newspapers. Same way, in case of radio, more participation of the experts, farmers, farm people can make the farm programmes more participative or demand oriented. Both the newspapers and radio should have taken more care regarding the time of disseminating farm information. Time period should be adjusted according to the time of different stages of cropping, so that it can help the farmers at right time. In terms of credibility of farm information, some particular areas revealed in the study should be given emphasis during dissemination of farm information. This study also highlighted that radio has higher credibility regarding farm information communication than that of newspapers. It is mentioned earlier that *All India Radio, Kolkata* has better understanding about the need and interest of the farm people. So, these areas need to be given more emphasis by the newspapers. As Government has taken some initiatives for the mass media support to the agricultural extension, these two media can be proved very useful in terms of farm information communication. Most of the newspapers are operated privately in India, so they should take more responsibilities of their own for the agricultural development in the country. Policy measures mentioned below can be extremely helpful to exploit these two media to their fullest potentialities, so that these two mass media can play a significant role in vertical development of agriculture on sustained basis. Then only food, feed and seed reserve in the country in future can be assured.

Policy Measures

The results of the study and its discussion enlighten different aspects of farm information communication through newspapers and radio. On the basis of these findings, some policy measures came out which are mentioned below:

- ☆ Number of farm or agriculture related information should be increased, both in newspapers and radio. Newspapers should allot a whole page on farm information in a given day in every week.
- ☆ Proper policy formulation is necessary to regulate some mass media, operated by private sector like newspapers, so that they incorporate more farm information in those media.
- ☆ Minimum amount of farm information should be stipulated for both newspapers and radio in relation to other information, which should be regulated by a governing authority.
- ☆ Both newspapers and radio should communicate farm information in more participatory mode, so

that the need and interest of farm people are paid proper attention.

- ☆ More location-specific information needs to be communicated to the farm people, so that they can make full use of the farm information.
- ☆ Credibility aspects of farm information should be taken care of by these two media. Farm information should be presented in such a way that farm people assume those farm information credible enough and subsequently can adopt them.
- ☆ Both newspapers and radio should develop 'Farm and Home Information Cell', which will operate the farm information effectively. This cell should possess technical experts from various disciplines of Agriculture, Horticulture, Home Science, Animal Husbandry, Rural Development, Nutrition etc. They can provide more input to farm information communication rather than general media personnel or officials.
- ☆ Newspapers and radio should develop an 'Evaluation and Research Team'. This team will be responsible for evaluating the efficacy of farm information in relation to need and interest of farm people, communicated by the concerned media. It should conduct media research in the field of farm information communication for better and effective communication of farm information.

References

Adigun, T.A. (2002). "Analysis of the publications of scientists at the West Africa Rice Development Association, Bouake, Cote d'Ivoire." *Quarterly Bulletin of IAALD*; **47(2)**: 43-46.

Angadi, J.G. and Swamy, B.S. (1995). "Credibility of Agricultural Assistants as perceived by farmers". *Karnataka Journal of Agricultural Sciences.* **8:** 1.10-14

Ani, A.O. and Kwaghe, P. V. (1997). "Sources of information on improved farm practices: a study of farmers in Umuahia Zone of Abia State, Nigeria". *Agro search.* **3(1/2):** 1-10

Ansaloni, F. (1986). "The role of the technical press in the diffusion of innovations in livestock production". *Questione-Agraria.* **22:** 175-194

Barcus, F.E. (1959). "Communications content ", Unpublished Ph.D. dissertation, University of Illinois.

Berelson, B. (1952). "Content Analysis in Communication Research", New York, Free Press.

Berelson, B. and Lazarsfeld, P. F. (1948). "The Analysis of Communication Content." Chicago: University of Chicago Press.

Berlo, D.K. (1960). "The process of communication: An introduction to theory and practice". New York, Holt, Rinehart and Winston.

Bharati, T. and Sohi, J.S. (2001). "Extension contacts and mass media exposure of the farmers in relation to the use of improved rice technology". *Himachal Journal of Agricultural Research.* **26(1/2):** 137-142.

Caldwell, A. M.; Bantle, S. K.; Cuperus, G. W.; Payton, M. E. and Berberet, R.C. (2000). "Media coverage of integrated pest management in major urban newspapers." *American Entomologist;* **46:** 1,56-60.

Carmines, E. G. and Zeller, R. A. (1979). "Reliability and Validity Assessment". SAGE publications India Pvt. Ltd. New Delhi, India.

Cartwright, D.P. (1953). "Analysis of qualitative material". In L. Festinger and D.

Katz (Eds.), Research Methods in the Behavioural Sciences. New York : Holt, Rinehart, and Winston. pp.421-470.

Census Data, 2001: Government of India

Chidera and Vajrathan (1973). "Communication and Rural Development". *Vidura.* **10:** 1

Conghnour, C.M. (1960). "The functioning of farmers characteristics in relation to contact with media and practice adoption". *Rural Sociology,* **25(3):** 286-288.

Cronbach, L.J. (1951). "Coefficient alpha and the internal structure of test". *Psychometrika* **16:** 297-334

Das, P.K. (1987). "Farm Journalists' Connection". Conference Paper: Conducted by Fertilizer Association of India (Eastern Region), Kolkata.

Das, P.K. and Sharma, J.K. (1998). "Credibility pattern of different sources of farm information". *Journal of the Agricultural Science Society of North East India.* **11:** 1, 121-123.

Dey, P.K. (1976). "A study of availability and utilization of sources of agricultural information in high and low urbanized villages". Unpublished Ph.D. thesis, Bidhan Chandra Krishi Viswavidyalaya, West Bengal.

Egbule, P. E. and Njoku, E. M. C. (2001). "Mass media support for adult education in agriculture in southern Nigeria". *Adult Education and Development.* **56:** 179-186.

Ensminger, D. (1957). "A guide to community development". Ministry of Community Development and Cooperation. Govt. of India. New Delhi.

Ernst, S.W. (1972). "Baseball or Brickbats: a content analysis of Community Development", *Journalism Quarterly,* Spring, **87.**

Fawole, O. P.; and Olowu, T. A. (2000). "Content analysis of women's agricultural activities in Nigerian newspapers, *Indian journal of Extension Education,* **36:** 1&2.

Flanagin, A.J. and Metzger, M.J. (2000). "Perceptions of Internet information credibility". *Journalism and Mass Communication Quarterly,* **77(3):** 515-540

Flesch, R. (1948). "A new readability yardstick". *Journal of Applied Psychology,* **32:** 221-233.

Flesch, R. (1974). "The art of readable writing". (Rev. Ed.), New York: Harper and Row.

Fox, J. (1963). "A Comprehension of Coothic Elite and Standard Elite Type Faces." *Economics*, **6:** 93-98.

Gazino, C. (1987). "News people's ideology and credibility debate". *Newspaper Research Journal*, **9(1):** 1-18

Gazino, C. and McGrath., K. (1986). "Measuring the concept of credibility". *Journalism Quarterly*, **63(3):** 451-462.

George, A. L. (1959). " Quantitative and qualitative approaches to content analysis." In I.D.S. Pool (Eds.), Trends on content analysis. Urbana: University of Illinois Press.

Gunning, R. (1952). "The technique of clear writing". McGraw Hill Berk Co., Inc.

Hagedorn, C and Allender, H.S (1998). "Public perceptions of transgenic in bio-control". *Ag Biotech News and Information.* **10:** 7, 217-221.

Hays, D.C. (1960). "Automatic content analysis." Santa Monica, CA. Rand Corporation.

Hovland, C.I. and Weiss, W. (1952). "The influence of source credibility on communication effectiveness". *Public Opinion Quarterly*, **15(4):** 635-650.

Hudeckova, H. and Lostak, M. (2004). "SAPARD: experiences and challenges for the future". *Zemedelska Ekonomika;* **50(4)**: 152-160.

Iker, H.P. (1974). "An historical note on the use of word frequency contiguities in content analysis". *Computers and Humanities*, **8:** 93-98

Indian Institute of Mass Communication (1968). "Agro-Information flow at the 'Village Level', New Delhi: Report of the Survey of Communication Sources in the Adoption Process of High Yielding varieties of crops".

Infante, D. A. (1980). "The constract validity of semantic scale for the measurement of source credibility". *Communication Quarterly,* **28(2):** 19-26

James, I.L. (1949). "The problem of validating content analysis in H.D. Lasswell, N. Leites *et al.* (Eds.), The language of politics : Studies in quantitative semantics", New York George Stewart, pp.55-82.

Janis, I.L. (1965). "The problem of validating content analysis". In Harold D. Lasseweell, Nathan Leites and Associates(Eds). Language of politics: Studies in quantitative semantics (pp.55-82). Cambridge: MIT Press. (Original work published in 1943).

Johnson, T. J. and Kaye, B. K. (1998). "Cruising is believing: Comparing Internet and traditional sources on media credibility measures". *Journalism and Mass Communication Quarterly,* **75(2):** 325-340.

Johnson, T. J. and Kaye, B. K. (2000). "Using is believing: The influence of reliance on the credibility of online political information among politically interested Internet users". *Journalism and Mass Communication Quarterly,* **77(4):** 865-879.

Jondhale, S.G. and Nandapurkar, G.G (1998). "An analytical study of utility perception of farm journals". *Journal of Maharashtra Agricultural Universities.* **23:** 3,294-297

Kalita, H.K. and Singh, M.K. (2003). "Credibility of information sources among tribal farmers". *Journal of Interacademicia.* **7 (4):** 457-460.

Kamath (1969). "Writing for Farm Families", New Delhi-Allied Publishers (P) Ltd., Bombay.

Kashem, M.A. (1999). "Farmers' contact with information sources in using farm technologies". *Bangladesh Journal Training and Development.* **12(1/2):** 61-68.

Kaur, D.K. and Mathur, P.N. (1981). "Format and cover page of an effective Hindi farm magazine". *Indian Journal of Extension Education.* **17:** I & 2.

Kayal, B.S. (1975). "A study of sources of information, readership and content in Bengali farm magazine 'Krishi Pragati' in West Bengal". Unpublished M.Sc. Thesis. Division of Agricultural Extension. IARI, New Delhi.

Kelsey, L.D. and Hearne, C.C. (1965). "Co-operative Extension Work", New York Comstock Publishing Associates.

Khandekar, P. (1975). "A study of effectiveness of a, farm magazine as related to different categories of readers". An unpublished M.Sc. Thesis. Division of Agricultural Extension, IARI, New Delhi.

Khandekar, P. and Mathur, P.N. (1975). "Effectiveness of a farm magazine 'Unnat Krishi'." *Indian Journal of Extension Education.* **12:** I & 2.

Kim, J. O. and Mueller, C. W. (1978). "Introduction to Factor Analysis; What It Is and How to Do It?" SAGE Publications India Pvt. Ltd. New Delhi, India.

Kiousis, S. (1999). "Public trust or mistrust? Perceptions of media credibility in the information age". Paper presented to the Mass Communication and Society Division, Association for Education in Journalism and Mass Communication, New Orleans.

Kota, T. (1983). "A study of readership and effectiveness of farm magazine 'Annadata' Unpublished M.Sc. Thesis". Bichpuri, Agra.

Krippendorff, K. (1992). "Recent developments in reliability analysis". Paper presented at the 42nd Annual Meeting of the International Communication Association, Miami, FL.

Krippendorff, K. (2004). "Content Analysis: An Introduction to its Methodology", Second Edition, CA: SAGE Publication.

Kumar, A. (1983). "Readership survey'of "Kisan Bharati" A farm magazine brought out by Pantnagar". Unpublished M.Sc. Thesis, G.B. P.U.A&T, Pantnagar.

Kumar, P. and Se, N. V. (2001). " Preferential modes of presentation of farm radio programme", *Indian Journal of Extension Education;* **37**: 1&2.

Ladebo, I. J.; Kassal, B. I. And Banjoko, O. C. (1997). "Effect of radio farm broadcasts on farmers knowledge of improved farm practices". *Journal of Extension System.* **13 (1/2):** 121-127.

Lasswell, H.D. (1941). "The World Attention Survey: An exploration of the possibilities of studying attention being given to the united states by newspapers abroad." *Public Opinion Quarterly,* **5:** 456-462.

Leites, N.C. and I. de S. Pool. (1942). "On content analysis", Washington, D.C. Library of Congress, Experimental Division for study of war-time communications, Document No.26.

Löbl, E. (1903). "Kultur und Presse." Leipzig, Germany: Duncker and Humbolt.

Maris, U. and Roe, K. (2004). "Automatic milking- a better understanding." Conference Proceedings, Lehystad, Netherlands; 41-45.

Markov, A.A. (1913). "Essai d'une recherché statisque sur le texte du roman "Eugene Onegin" illustrant la liaison des epreuves en chaine (Russian)". *Bulletin de l'Academie Imperiale des Sciences de St. Petersbourg.* **6(7):** 153-162.

Marks, L. A.; Kalaitzandonakes, N.; Allison, K. and Zakharova, L. (2003). "Media coverage of agro biotechnology: did the butterfly have an effect?" *Journal of Agribusiness;* **21(1)**: 1-20.

Mathews, B.C. (1910). "A study of a New York daily." *Independent,* **68:** 82-86.

Mathur, I.N., Singh, K.N. and Das, K.K. (1978). "Readership pattern, nature of content and sources of information of farm periodicals in India". *Indian Journal of Extension Education,* **14:** 3 & 4, pp.39-44.

Mayer, P. (1988). "Defining and measuring credibility of newspapers: Developing an index". *Journalism Quarterly,* **65(3):** 567-574.

McHenry, H. (1996). "Farming and environmental discourses: a study of the depiction of environmental issues in a German farming newspaper". *Journal of Rural Studies.* **12:** 4, 375-386.

Mishra, B.P. and Tripathy, A. (1991) "The extent of exposure of farm women to different sources of information and the constraints regarding low and non-utilization of the sources. *Orissa Journal of Agricultural Research.* **4(3/4):** 221-223.

Mulder, R. (1980). "Media credibility: A use gratification approach". *Journalism Quarterly,* **57(3):** 474-477.

Mulder, R. (1981). "A long-linear analysis of media credibility". *Journalism Quarterly,* **58(4):** 635-638.

Naccarato, J. L., and Kimberly, A. N. (1998). "Content Analysis as a Predictive Methodology : Recall, readership, and Evaluations of Business-to-Business Print Advertising". *Journal of Advertising Research.* **38(3):** 19-30.

Nagabhushanam, K. and Nataraju, M. S. (2000). "Content analysis of feature articles in Kannada Daily". *Mysore Journal of Agricultural Sciences;* **34(2):** 157-160.

Namenwrith, J.Z. and Weber, R.P. (1987). "Dynamics of culture." Winchester, MA: Allen and Unwin.

Nanjappa, D. and Ganapathy, K.R. (1987). "Contest analysis of agriculture as information in selected Kanada Dailies". *Indian Journal of Extension Education,* **23:** 192, 20-23.

Nickerson, C. J. and Hellerstein, D. (2003). "Protecting rural amenities through farmland preservation programmes." *Agricultural and Resource Economics Review;* **32(1)**: 129-144.

Ognianova, E. (1998). "The value of journalistic identity on the World Wide Web". Paper presented to the Mass Communication and Society Division, Association for

Education in Journalism and Mass Communication, Baltimore.

Oladeji, J.O; and Olowu, T.A. (2003). "Information cues of agricultural advertisements in Nigerian daily newspapers." *Moor Journal of Agricultural Research*; **4(2):** 260-265.

Omotayo, A. M.; Chikwendu, D. O., Zaria, M. B.; Yusuf, J. B. and Omenesa, Z. E. (1997). "Effectiveness of radio in Nigeria in dissemination of information on improved farming practices". *Journal of Extension System.* **13(1/2):** 103-120.

Osgood, C. E. (1959). " The representation model and relevant research methods. In Ithiel de Sola Pool (Eds.), Trends on content analysis" (pp.33-38). Urbana: University of Illinois Press.

Oyegbami, A. and Fabusoro, E. (2003). "The use of radio and television as sources of agricultural information among poultry farmers in Egbeda local government area of Oyo State, Nigeria."; *Moor Journal of Agricultural Research*; **4(1)**: 164-169.

Pal, P.P.; Sundarambal, P.; Bihari, B.; Kumar, R. and Prakash, N. C. (1999). "Information seeking behaviour of the tribal farmers of Meghalaya". *Indian Journal of Hill Farming.* **12(1/2):** 124-125.

Pareek, U. and Trivedi, G. (1964). "Manual of socio-economic status scale (Rural)". Delhi, Manasayan.

Patel, M. M.; Parmer, P. S. and Dubey, M. C. (1995). "Credibility pattern of different sources of farm information." *Maharashtra Journal of Extension Education*; **15:** 33-34.

Patel, S.M. and Patel, H.N. (1972). "Study and evaluation of "Khedat Patrika" published by Information Unit of the Development of Agriculture, Gujarat State. *Indian Journal of Extension Education,* **8:** I & 2, pp.33-36.

Peterson, T. (1964). "Magazine in the Twentieth Century", Urbana; University of Illinois Press.

Pool, I. de S (1959). "Trends in content analysis". Urbana : University of Illinois Press.

Popping, R. (1997). "Computer programme for the analysis of text and transcript." In Carl. W. Roberts (Eds.), Text analysis for the social sciences: Methods for drawing statistical inferences from text and transcripts (pp.209-221). Mahwah, NJ: Lawrencee Erlbaum.

Potter, W.J. and Levine-Donnerstein, D. (1999). "Rethinking reliability and validity in content analysis". *Journal of Applied Communication Research,* **27:** 258-284

Rahman, M. A.; Parveen, S. and Halim, A. (1999). "Credibility of information sources in disseminating HYV *Boro* rice technology perceived by the block supervisors". *Annals of Bangladesh Agriculture.* **8(2):** 105-110.

Rai, H.N. (1965). "Diffusion of information and farmers' response in relation to an Improved Farm Practice (Hybrid Maize)". *Indian Journal of Extension Education.* **5:** 140.

Ray, G.L. (1967). "A study of some Agricultural Practices in Bangarupalam Panchayat Samity Block, Chitoor District, Andhra Pradesh". Unpublished M.Sc. Thesis. Andhra Pradesh Agricultural University, Hyderabad (India).

Reddy, P. R. and Reddy, S. J. (1997). "Listening behaviour of farmers towards farm broadcasts in Nizamabad district of Andhra Pradesh - a path analysis". *Journal of Research, ANGRAU.* **25(2):** 31-34.

Resner, A. (2003). "Newspaper construction of a moral farmer." *Rural sociology,* **68:** 1, 46-63.

Riffe, D., Lacy, S. and Drager, M.W. (1996). "Sample Size in Content Analysis of Weekly News Magazines". *Journalism and Mass Communication Quarterly.* **73(3):** 635-644.

Riley, J. W. and Riley M. (1959). "Mass Communication and the Social System" In Robert, K. Metron and others (Eds.). Sociology Today: Problems and Prospects. New York: Basic Books.

Rimmer, T. and Weaver, D. (1987). "Different questions, different answers? Media use and media credibility". *Journalism Quarterly,* **64(1):** 28-36.

Rodriguez, L.A.; Schwab, C.V.; Peterson, J.W. and Miller, L.J. (1997). "Safe farm: the impact of a lowa public Information Campaign". *Journal of Agricultural Safety and Health.* **3(2):** 109-123.

Rogers, E.M. and Beal, G.M. (1958). "The Importance of personal influence in the adoption of Technological Changes". *Social Forces.* **36:** 329-335.

Roy, S.; Verma, N. S.; Laharia, S. N. and Sarangi, A. (2003). "Content and coverage of farm news in leading Bengali dailies". *Indian Journal of Extension Education;* **39**: 3&4.

Sarkar, A. (1997). "Impact of mass communication on rural people in relation to agricultural information". *Environment and Ecology.* **15(4):** 862-866.

Sasidhar, P.V.K.; Rao, B. S. and Reddy, S.R. (1999). "An analysis of the programmes of farm and home broadcast". *Journal of Research-ANGRAU;* **27(4):** 87-91.

Schwarz, J. and Doluschitz, R. (1997). "Need for new media by farm businesses-results of an exploratory empirical study". *Zeitschrift fur Agrarinfermatik.* **5(5):** 100-107.

Schweiger (1998). "Wer glaubt dem World Wide Web? Ein experiment zur glabwurdigkeit von nachrichten in tageszeitungen und im World Wide Web, a in Rossler, P. (ed.)". *Online-kommunikation beitrage zu und wirkung.* pp.123–145, Opladen: Wesduetscher Verlag.

Schweiger, W. (2000). "Media credibility of experience or image? A survey on the credibility of the world wide web in Germany in comparison to other media". *European Journal of Communication.***15:** 37-59.

Sebeok, T. A. and Zeps, V.J. (1958). "An analysis of structured content with application of electronic computer research in psycholinguistics". *Language and Speech,* **1:** 181-193

Silva, D.N.de and Garforth, C.J. (1997). " The effectiveness of different radio programme format for the dissemination of information on the safe use of insecticides in paddy cultivation in Mohaweli system–C of Srilanka". *Tropical Agricultural Research.* **9:** 302-316.

Singh, N. G. N. (2001). "Content analysis of farm broadcast programme, Imphal. *Indian Journal of Extension Education;* **37**: 3&4.

Singh, Y.P. and Nirwal, R.S. (1974). "Audience analysis for using words". *Community Development and Panchayat Raj Digest,* **6:** 1.

Smythe, D.W. (1952). "Some observations of communications theory". *Audiovisual Communication Review,* **2:** 24-37.

Somasundaram, T. and Jagannath, L. (1974). "Readability in News Letter". *Indian Journal of Extension Education.* **10:** 1 & 2.

Speed, G.J. (1893). "Do newspapers now give the news?" Forum, **15: 705**-711.

Stefflre, V. (1965). "Simulation of people's behaviour toward new objects and events". *American Behavioural Scientist,* **8:** 12-15

Stone, P.J. (1975). "Report on the workshop on content Analysis in the Social Sciences, Pisa, 1974". *Social Science Information,* **14:** 107-111

Stone, P.J., Dunphy, D.C., Smith, M.S. and Ogilvie, D. M. (1966). "The General Inquirer: A Computer Approach to Content Analysis." Cambridge: MIT Press.

Street, A. T. (1909). "The truth about newspapers." *Chicago Tribune.*

Sunder, S. (1996). "Do quotes affect perception of online news stories?" Paper presented to the Communication Technology and Policy Division, Association for Education in Journalism and Mass Communication, Anaheim, Calif.

Sunder, S. (1999). "Exploring receivers' criteria for perception of print and online news". *Journalism and Mass Communication Quarterly,* **76(2):** 373-386.

Sur, A. (1994). "Bangla & Bangalir Bibartan (An Ethno-Cultural History of Bengal)" Published by Sahityalok, Kolkata,

Swant, G.K. (1982). "An analytical study of printed fertilizer advertisements and their communicability and effectiveness as perceived by farmers of Maharastra". Unpublished Ph.D. Thesis, IARI, New Delhi.

Swant, G.K. and Shendre, N.K. (1987). "Coverage of agricultural information in a leading Marathi newspapers 'Sakal' from Maharashtra". *Journal of Maharashtra Agricultural Universities.* **12:** 2, 232-235.

Taylor, W. (1953). "Close Procedure, a new tool for measuring readability". *Journalism Quarterly,* **30:** 415-433.

Vannirajan, P.; Ravichandran, V. and Velusamy,R. (2000). "Preferences of farmer readers towards various contents of the farm information ". *Madras Agricultural Journal.* **86(7/9):** 375-378.

Velusamy, R. and Seetharaman, R.N. (2000). "Content analysis of farm information in selected dailies." *Madras Agricultural Journal.* **87**: 7-9,408-410.

Vyas, R. V. (2002) "Educational Radio in India". *Turkish Online Journal of Distance Education-TOJDE* ; **3:** 3

Wanta, W. and Hu, Y. (1994). "The effects of credibility, reliance and exposure on media agenda-setting: A path analysis model". *Journalism Quarterly,* **71(1):** 90-98.

Weber, M. (1911). "Geschaftsbericht" in verhandlungen des Ersten Deutschen Soziologietages." In Schrift der Deutschen Gessellschaft fur Soziologie (pp.39-62) Vom **19-22.** Oktober 1910 in Frankfurt A.M.

Weber, R. P. (1990). "Basic Content Analysis". Second Edition. CA: Sage Publications, New Delhi.

www.accenter.com/aboutthecenter/glossary.asp

www.allindiaradio.org/

www.amsglossary.allenpress.com/glossary/browse

www.angelfire.com/trek/puknews/refsnp.html

www.attorneykennugent.com/library/d.html

www.bfa.sdsu.edu/ehs/glossary.htm

www.borealforest.org/nwgloss4.htm

www.calforests.org/glossary.html

www.ccs.mit.edu/21c/iokey.html

www.cdc.gov/nccdphp/drh/epi_gloss.htm

www.cdc.gov/tobacco/evaluation_manual/glossary.html

www.collections.ic.gc.ca/ghost/glossary.html

www.dairynet.com/kids/gloss.html

www.e11th- hour.org/resources/backgrounders/environmental.glossary.a.html

www.ebfarm.com/Organic/Glossary.aspx

www.en.wikipedia.org/wiki/Advertisement

www.en.wikipedia.org/wiki/Agricultural_machinery

www.en.wikipedia.org/wiki/Agricultural_policy

www.en.wikipedia.org/wiki/Calamity

www.en.wikipedia.org/wiki/Credibility

www.en.wikipedia.org/wiki/Farm

www.en.wikipedia.org/wiki/Fishery

www.en.wikipedia.org/wiki/Information

www.en.wikipedia.org/wiki/Interview

www.en.wikipedia.org/wiki/Marketing

www.en.wikipedia.org/wiki/News

www.en.wikipedia.org/wiki/Presentation

www.en.wikipedia.org/wiki/Radio_play

www.en.wikipedia.org/wiki/Repetition

www.en.wikipedia.org/wiki/Women

www.fao.org/docrep/003/X3910E/X3910E22.htm

www.fcps.k12.va.us/StratfordLandingES/Ecology/mpages/glossary_qz.htm

www.fhwa.dot.gov/realestate/rights/defin.html

www.ggca4thgrade.tripod.com/id10.html

www.globalsecurity.org/wmd/library/policy/army/fm/3-6/3-6gl.htm

www.grants.nih.gov/grants/funding/phs398/instructions2/p3_definitions.htm

www.gtscompanies.com/glosscomp.html

www.gulflink.osd.mil/medsearch/glossary/glossary_b.shtml

www.health.nsw.gov.au/public-health/health-promotion/abouthp/glossary.html

www.highered.mcgraw-hill.com/sites/007256296x/student_view0/glossary.html

www.hindubooks.org/temples/bengal/birbhum/index.htm

www.inspection.gc.ca/english/corpaffr/publications/riscomm/riscomm_appe.shtml

www.intra.ccgh.com/AutomatedCAMH/gl.html

www.ipmalmanac.com/glossary/index.asp

www.it.bton.ac.uk/staff/je/adacraft/glossary.htm

www.learn-english-today.com/business-english/advertising-vocabulary.html

www.lineadecreditoambiental.org/html/glossary.html

www.mapp.naccho.org/mapp_glossary.asp

www.mvp.cfee.org/en/glossary.html

www.nscb.gov.ph/ru12/DEFINE/DEF-AGRI.HTM

www.oahutrans2k.com/info/glossary/T.htm

www.openheartlogic.org/opine.cgi

www.orafaq.com/glossary/faqglosi.htm

www.phy.ohiou.edu/academicpro/indepen/One/glos.htm

www.preemptinc.com/definitions.html

www.preferredlendingpartners.org/programterms.htm

www.rho.org/html/glossary.html

www.student.biology.arizona.edu/honors98/group14/glossary.html

www.terena.nl/tech/task-forces/tf-csirt/iodef/docs/i-taxonomy_terms.html

www.thegardenhelper.com/dictionary.html

www.thequalityportal.com/glossary/f.htm

www.ucblibraries.colorado.edu/about/glossary.htm

www.unesco.org/education/tlsf/theme_c/mod13/
www.worldbank.org/depweb/english/modules/glossary.htm

www.wordnet.princeton.edu/perl/webwn

www.xdcr.com/UT_terms.html

Yiping, D.; Tsunashima, F. and Ozawa, W. (2002). "The movement of agricultural journalism and farming readers after the war. *Journal of Rural Economics.* **74:** 81-85.

Zhu, Y. X.; Zhou, J. G.; Zhu, Y. X. and Zhou, J.G. (2003). " Searching coordinating and analyzing of tea literature published in China." *Journal of Zhejiang University of Agriculture and Life Sciences;* **29:** 6, 689-693.

www.ingramcontent.com/pod-product-compliance
Lightning Source LLC
Chambersburg PA
CBHW071222230926
51849CB00001B/5
9789351241027